AF587996

Computer Science, Technology and Applications

Computer Science, Technology and Applications

Speech Recognition Technology and Applications
Vasile-Florian Păiş (Editor)
2022. ISBN: 978-1-68507-929-1 (Hardcover)
2022. ISBN: 979-8-88697-179-8 (eBook)

Internet of Everything: Smart Sensing Technologies
T. Kavitha, PhD, V. Ajantha Devi, PhD,
S. Neelavathy Pari, PhD
and Sakkaravarthi Ramanathan, PhD
(Editors)
2022. ISBN: 978-1-68507-865-2 (Hardcover)
2022. ISBN: 978-1-68507-943-7 (eBook)

A Beginner's Guide to Virtual Reality (VR) Modeling in Healthcare Applications with Blender
Ho Lun Ho, Ka Yin Chau, PhD,
Yan Wan and Yuk Ming Tang, PhD (Editors)
2022. ISBN: 978-1-68507-811-9 (Softcover)
2022. ISBN: 978-1-68507-945-1 (eBook)

Applying an Advanced Information Search and Retrieval Model in Organisations: Research and Opportunities
Maria del Carmen Cruz Gil
2022. ISBN: 978-1-68507-560-6 (Softcover)
2022. ISBN: 978-1-68507-914-7 (eBook)

Neural Network Control of Vehicles: Modeling and Simulation
Igor Astrov
2022. ISBN: 978-1-68507-757-0 (Hardcover)
2022. ISBN: 978-1-68507-916-1 (eBook)

More information about this series can be found at
https://novapublishers.com/product-category/series/computer-science-technology-and-applications/

Yogesh Kumar Chauhan,
Ranjan Kumar Behera
and Asheesh K. Singh
Editors

Applied Artificial Intelligence (AI) to Green Power Technology

Copyright © 2022 by Nova Science Publishers, Inc.
DOI: https://doi.org/10.52305/ISQD2111

All rights reserved. No part of this book may be reproduced, stored in a retrieval system or transmitted in any form or by any means: electronic, electrostatic, magnetic, tape, mechanical photocopying, recording or otherwise without the written permission of the Publisher.

We have partnered with Copyright Clearance Center to make it easy for you to obtain permissions to reuse content from this publication. Simply navigate to this publication's page on Nova's website and locate the "Get Permission" button below the title description. This button is linked directly to the title's permission page on copyright.com. Alternatively, you can visit copyright.com and search by title, ISBN, or ISSN.

For further questions about using the service on copyright.com, please contact:
Copyright Clearance Center
Phone: +1-(978) 750-8400 Fax: +1-(978) 750-4470 E-mail: info@copyright.com.

NOTICE TO THE READER

The Publisher has taken reasonable care in the preparation of this book, but makes no expressed or implied warranty of any kind and assumes no responsibility for any errors or omissions. No liability is assumed for incidental or consequential damages in connection with or arising out of information contained in this book. The Publisher shall not be liable for any special, consequential, or exemplary damages resulting, in whole or in part, from the readers' use of, or reliance upon, this material. Any parts of this book based on government reports are so indicated and copyright is claimed for those parts to the extent applicable to compilations of such works.

Independent verification should be sought for any data, advice or recommendations contained in this book. In addition, no responsibility is assumed by the Publisher for any injury and/or damage to persons or property arising from any methods, products, instructions, ideas or otherwise contained in this publication.

This publication is designed to provide accurate and authoritative information with regard to the subject matter covered herein. It is sold with the clear understanding that the Publisher is not engaged in rendering legal or any other professional services. If legal or any other expert assistance is required, the services of a competent person should be sought. FROM A DECLARATION OF PARTICIPANTS JOINTLY ADOPTED BY A COMMITTEE OF THE AMERICAN BAR ASSOCIATION AND A COMMITTEE OF PUBLISHERS.

Additional color graphics may be available in the e-book version of this book.

Library of Congress Cataloging-in-Publication Data

ISBN: 979-8-88697-131-6

Published by Nova Science Publishers, Inc. † New York

Contents

Preface

Recently, governments all over the world have focused on reducing their carbon footprint, generating a deep impact on policies of government, apart from public and private sectors. Now, the world leaders have agreed on sustainable development, technology, and energy to improve life on earth with less damage to the environment and ecology.

Nowadays, artificial intelligence (AI) has a wider reach among all areas of engineering and technology for enhancing productivity, efficiency and performance. Green power sources are one such area. The major green power sources are solar, wind, small hydro, natural and hydrogen gas, etc. Most of these sources have issues of intermittency and non-linear relationship with input and output variables. So, there is a need to optimize the performance of these resources to improve their outlook for widespread use and acceptability.

Among the latest AI techniques, fuzzy logic and nature-inspired algorithms are leading from the front. These techniques are applied to improve the performance of renewable energy systems, green power technologies such as solar PV system/wind energy conversion system etc. and control of power electronics interfaces for renewable energy systems.

The aim of this book is create awareness and generate interest among UG/PG students, research scholars, engineers, scientists, and regulators and policy makers of government and the public and private sectors. The book will have wider reach in audience as it is application area of AI to green power technologies.

Contributors to this work are some of the most eminent personalities and researchers in the field. The latest issues and challenges in green power technologies will be addressed. The newly developed AI tools and techniques will be proposed and implemented for the solution in this area. Here, the work embodied in this book will be focused on the latest green power technologies and their improvement with AI tools and techniques, which will have novel contributions and will be presented uniquely. The newly developed AI tools and techniques will be proposed and implemented for the solution in this area.

The editors feel that the book Applied Artificial Intelligence (AI) to Green Power Technology will provide a satisfying experience for the audience. They will be immensely benefitted by the diversified coverage and in-depth analysis.

Yogesh Kumar Chauhan, Ranjan Kumar Behera, Asheesh K. Singh
Editors

Acknowledgments

It is a matter of great privilege and pleasure to acknowledge all the persons who have contributed to this edited book entitled Applied Artificial Intelligence (AI) to Green Power Technology, published by Nova Science Publishers.

First of all, my heartfelt gratitude goes to the co-editors (Prof. Ranjan Kumar Behera, EED, IIT Patna and Prof. Asheesh Kumar Singh, EED, MNNIT Prayagraj), who have extended all help and support from the inception of the idea to the embodiment of this book.

All my thanks go to the authors who have contributed to this book, whose chapters were accepted after a thorough review process.

My special thanks go to all the esteemed reviewers, for their effort and for sparing time from their busy schedules to review the chapters assigned to them.

My heartfelt thanks go to all the efforts made by Nova Science Publishers for accepting the book proposal, extending their help from time to time and always standing with me throughout the whole period to complete this task.

I extend my thanks to all persons, institutions, organizations etc. who have contributed directly or indirectly during preparation of this manuscript.

Dr. Yogesh Kumar Chauhan
Editor

Chapter 1

Energy Management and Artificial Intelligence

Subhash Yadav[*]**, Pradeep Kumar**[†]
and Ashwani Kumar[‡]
National Institute of Technology, Kurukshetra, India

Abstract

Conventional energy sources availability is limited and has several environmental concerns associated with it. Renewable energy sources (RESs) provide clean energy to address the concerns raised by conventional energy sources. Energy management helps solve the issue arising due to the use of conventional energy sources and their impact on the environment. AI uses the data to ease the task of solving the problem of energy management in modern energy management systems. This chapter provides a brief overview of the application of AI for Energy Management purposes. It provides a framework and critical issues associated with this application.

Keywords: energy management, artificial intelligence, renewable energy sources

[*]Corresponding Author's Email: subhash05754@gmail.com.
[†]Corresponding Author's Email: pradeepkumar@ieee.org.
[‡]Corresponding Author's Email: ashwani.k.sharma@nitkkr.ac.in.

In: Applied Artificial Intelligence (AI) to Green Power Technology
Editors: Yogesh Kumar Chauhan, Ranjan Kumar Behera and Asheesh K. Singh
ISBN: 979-8-88697-131-6
© 2022 Nova Science Publishers, Inc.

Introduction

Electricity has become an essential part of human life. Humans use electrical energy to complete a lot of their daily human activities (Armaroli and Vincenzo, 2007). The electricity generated through different processes is converted into other useful forms of energy at the end. With rising dependence on electricity, the electricity demand is increasing day by day (Asif and Tariq, 2007), (Leal et al., 2022). Figure 1 shows the major energy consumers and their demand growth over the past few years ("As of February 16, 2022, World Energy Climate Statics-Yearbook 2021"). Most of these countries are developed countries or developing at a fast rate. Earlier, the synchronous generators generated electrical energy using the conventional generation. The conventional energy sources have limited reserves with climate concerns. Thus, the world is moving towards renewable energy sources (RESs). The RESs provide clean energy to address the concerns raised by conventional energy sources. The RESs have further helped increase the electrification rate ("As of February 16, 2022, IEA 2022"). The electrification rate for countries with the highest electrification rate is shown in Figure 2 ("As of February 16, 2022, World Energy Climate Statics-Yearbook 2021 Electricity Consumption Rate"). These figures show that the countries in both the figures are different; the countries having high demand does not have a high electrification rate. This can be attributed to the technological shift in electricity generation. Earlier, the conventional energy sources driving the synchronous generator were distributed throughout the world in a non-uniform manner. Thus the energy demand was also distributed in the same manner. However, with the developments in RESs generation technology, energy supply and demand patterns are shifting, resulting in different electrification rates and electricity demand. With these shifts in generation demand, the utilities need to address some more challenges related to the energy sector (Cheng et al., 2021).

- With increasing dependence on electrical energy and rising demand,
- Low overall system efficiency,
- Changing supply and demand patterns, and
- Lack of analytics needed for optimal energy management.

These challenges are present in almost all electric utilities across the globe. Amongst these issues, the low system efficiency is particularly problematic, as low efficiency means a large amount of power is neither measured nor billed, resulting in (Makala and Tonci, 2020).

- Losses,
- Greater CO_2 emissions,
- Consumers have little incentive to use the energy they don't pay for rationally.

Energy management is an essential tool that helps address the challenges associated with the demand and supply of electricity. Energy management has existed as a critical tool for industries and utilities to improve system efficiency and manage resources (Doty and Wayne, 2004). With the changing energy scenario, the importance of energy management has further increased. The problem has transformed from system efficiency improvement to resource management. The energy management in microgrids refers to an information and control system-based functionality ensuring minimum cost operation (Yimy et al., 2019). The energy management has been performed using several optimization algorithms. Recent developments in Artificial Intelligence (AI) have made it possible to use energy management using AI ("As of March 01, 2022, The Role of AI Technology in improving the renewable energy sector"). This chapter presents energy management using AI.

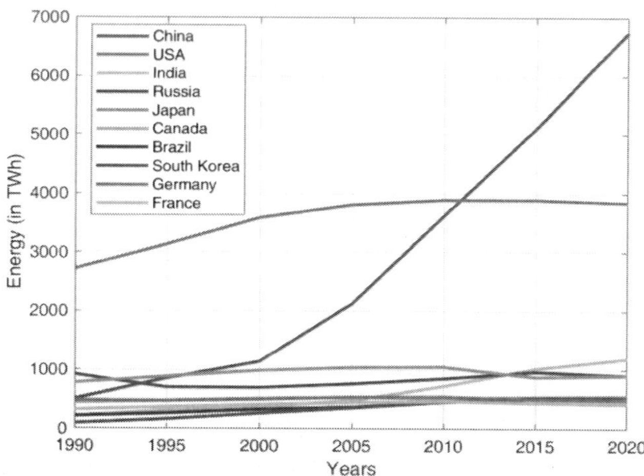

Figure 1. Major electricity consumers throughout the world and their energy consumption.

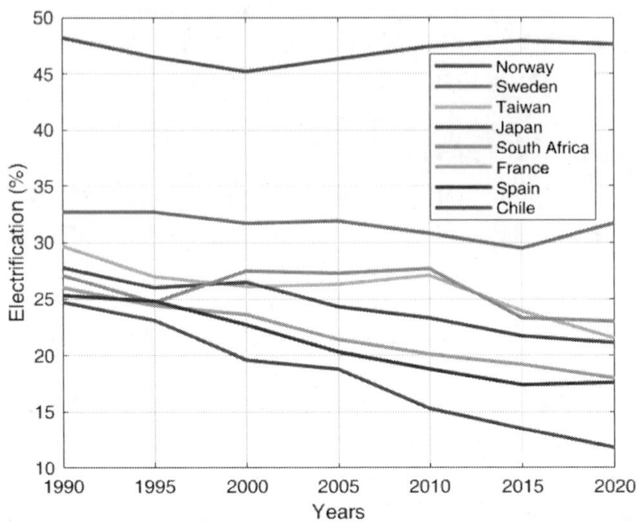

Figure 2. Rate of electrification in some leading countries throughout the world.

The different sections in the paper are as follows: Section-2 is the importance of energy management and its process. Section-3 discusses the electric grid and energy management. Section-4 and Section-5 discuss Artificial Intelligence (AI) and AI for Energy Management.

Energy Management

Overview

The modern day energy system has several energy sources, including conventional and non-conventional sources or RESs. The conventional energy sources are limited in nature and pollute the environment. From the operational point of view, their dispatchable nature allows for scheduling the generation as per requirement (Yaosuo, Liuchen, and Julian, 2007). These sources are also unidirectional, located at a centralized location, from where the power flows towards the load. On the other hand, the RESs are located on the generation and distribution sides. These sources are diversified and decentralized, i.e., located across generation, transmission, and distribution, making them bidirectional (Alanne and Arto 2006). The bidirectional power flow is difficult to handle using existing transmission and distribution structures designed to handle unidirectional power flow. The weather dependence of the RES makes the source non-dispatchable in nature. To make them dispatachable, the use of storage is essential. It helps RESs to make it match the load demand. RESs and batteries make the system complex with several operational and management issues. Whether conventional or non-conventional sources, management of these resources, while minimizing the cost, as energy management is the key as it is helpful in (Doty and Wayne, 2004).

- National good
- Management of resources.
- Reduction in energy waste.
- Manage the increasing demand.

Objectives

Thus, energy management adjusts and optimizes the energy source to achieve the following objectives (i) match the generation and load, (ii) schedule the different energy sources, and (iii) reduce the total energy production cost while maximizing the profits (Sebastian and Shafiuzzaman, 2018). Among these objectives, profit maximization and cost minimization are the primary objectives. The monetary benefits achieved through this form the motivation for the people to undergo energy management. Some other sub-objectives desired for energy management are (Doty and Wayne, 2004).

- Reduce energy wastage and reduce energy costs.
- Reduce green house gas emissions and improve air quality.
- Improve information flow related to the energy issues.
- Effective energy monitoring, reporting, and management for improved energy usage.
- Improve Returns from energy investments.
- Develop an inclusive program where every stakeholder can participate in energy management.
- Reduce the impacts of an interruption in energy supplies.

To achieve these objectives, understanding the energy management process is critical.

Energy Management Process

The objectives presented above are achieved using the energy management process shown in Figure 3. Energy management is a target based approach, where the targets are set for future energy consumption using past energy consumption patterns. The assessment of future and base loads is critical to plan for future energy requirements. This future growth also makes energy management iterative, implemented either continuously or after a regular interval. Thus, it becomes an energy policy matter to decide the frequency and level of energy management permitted in a facility. Before performing the energy management, the following quantities are collected (Doty and Wayne, 2004).

- Dependent and independent variable.
- Operating hours.
- Temperature profile of the surrounding environment.
- Energy consumption data.

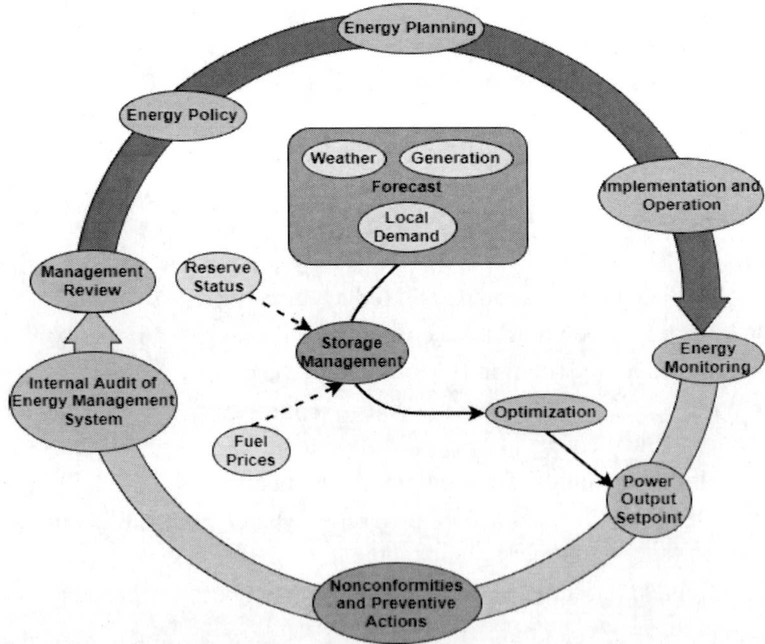

Figure 3. Energy management process.

The variables are the quantities of the energy management systems. Ones affecting energy consumption are dependent variables, whereas others are independent variables. In a RES based microgrid, the crucial variables to consider are the weather, load, and generation profiles. These parameters are also forecasted to get the future energy generation. The fuel prices and energy storage status are also analyzed before the optimization. Optimization is performed using this data to generate an optimal energy management schedule with the desired objectives discussed in the sub-section above. The optimized energy management output is compared with the energy monitoring data over the years to check whether the targets are achieved or not. This information is also helpful in setting the energy management objectives of the future. The non-conformities are identified using the information and preventive actions that are to be taken to achieve the target. These non-conformities and the process through which they are obtained are audited before implementation. These outcomes are then reported to the management for review. This performance reporting is also vital to monitor the overall cost and keep the motivation intact for all the stakeholders involved.

Moreover, this reporting also verifies the process and targets achieved. After that, the management reviews these actions as per the energy policy at the energy planning level. In the next iteration of energy management, the results are reflected in the targets and energy monitoring levels. Another critical parameter to keep the motivation for the whole process is to transfer the cost-benefit to all the stakeholders.

Electric Grid and Energy Management

The energy management process described in the section above shows that the data is essential to the whole operation. The data is collected and processed at different levels. Energy auditing is also performed based on the information obtained through measurement data. Summarily, the whole energy management process is dependent on,

- Energy Audit
- Data Analytics and
- Information Flow.

The extraction and utilization of this information require a lot of advanced metering infrastructure (Ramyar et al., 2014). The conventional electric grid was low in coordination, planning, and reliability. Moreover, the infrastructure is also aging; making it difficult to perform reliably, and fuel price volatility has further limited the capabilities of the grid operators to provide low costs to the consumers. Moreover, the customers were primarily left out of the planning and operation process. They were on the receiving end of the system, responding to the changes occurring in the system. This shows the importance of communication in the whole system.

The smart grid allows communication to be an integral part of the system. It enables the communication between different devices and stakeholders to make the electric grid participatory. It is considered as a key component to integrate the RES into the smart grid. To achieve this, metering infrastructure is also essential. The five-layer structure for the smart grids is shown in Figure 4 (Abouzar, Edoardo, and Luca, 2018). Amongst these layers, the energy management function is performed at the functional layer. The functional layer deals with the different functions performed over the grid using the information layer. The information layer gathers the information from the communication system. Thus, it fulfills the requirements for energy management as laid down above, i.e., data analytics and information flow. The output of the functional layer is transferred to the business layer. The business layer deal with the business relevant information from the utility point of view. Thus, it shows that the smart grid is one of the essential building blocks for incorporating the Energy management system.

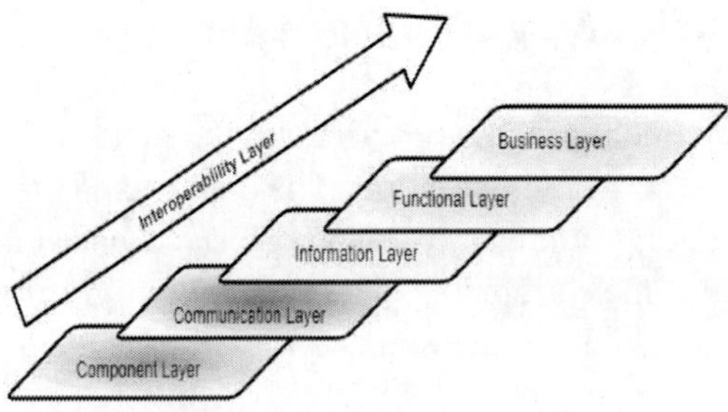

Figure 4. Five-layer structure of smart grids.

A significant problem related to the energy management system at the functional layers is handling the amount of information. The communication layer transfers the measured data from the component layer to the information layer. This leads to the problem of big data analytics. This should be handled reliably, as mishandling of the information can lead to incorrect decisions related to the system. Data analytics is a helpful tool for performing this task. However, human interventions can lead to the loss of information from the system. This is the point where automated data analytics comes into the picture. Artificial Intelligence (AI) is considered a reliable tool to handle the information and extract the relevant information for the operators to decide further action. The AI can also help with decision-making; however, it is subject to the policy of the concerned system operation. The details of how AI helps handle the information are discussed next.

Artificial Intelligence

Artificial Intelligence means transferring human Intelligence to computing devices (T.Steven, 2015). AI has existed for a long time. However, AI has always been cited as a risky affair for the human race due to the power provided to machines (Andreas and Michael, 2020). The AI includes all the machines learning new things from experience, adapting to the situation, and decision making. The different components of AI are shown in Figure 5 (Sara et al., 2021) (Manivannan et al., 2022). The Figure shows the broad categories of AI. Each category has different algorithms.

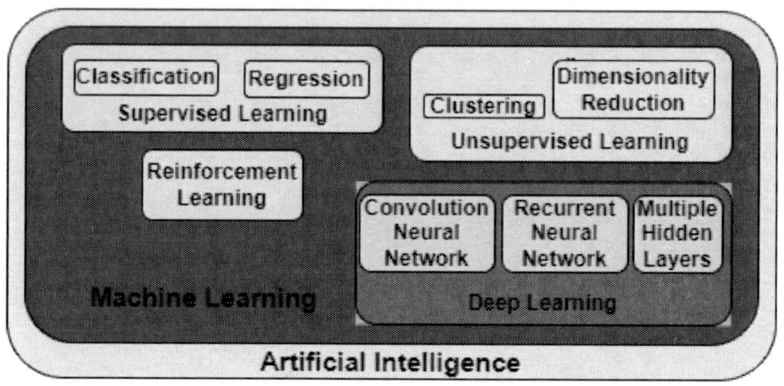

Figure 5. Overview of the AI system.

As discussed in Section-3, the data in the smart electric grids is critical for the operation of the grid. The data analytic framework helps to extract the information. AI helps perform this task of analysis and decision-making. It uses automated algorithms to process the data and extract the information. Also, the complete system learns itself and keeps updating the knowledge base. Due to this reason, AI has been considered a suitable choice for increasing the situation awareness about the grid, automation, and control of different power apparatus (Syed and Bong, 2020). Some other critical applications of AI for electrical engineering (Makala and Toni, 2020).

- Fault prediction
- Image processing facilitated power apparatus maintenance
- Decision making
- Disaster recovery.
- Loss prevention due to informal connections.

AI for Energy Management

As discussed above, AI helps increase the situational awareness of the grid and decision-making. The energy management of one such process used in the network. The algorithm for energy management using AI is shown in Figure 6. The energy management process is already shown in Figure 3.

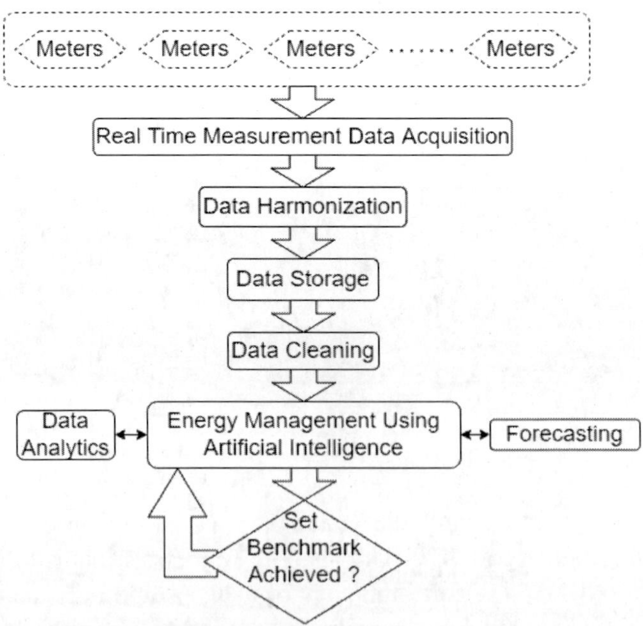

Figure 6. Information flow for applying AI to energy management.

With AI, the process remains the same; however, AI is helpful in learning and decision-making. Figure 6 shows that the real-time data from the meters is acquired using data acquisition. The data is harmonized after acquisition. The harmonization process allows the data from different meters to be combined in an informative manner, making its use easy. The harmonized data is then stored and filtered. The clean data obtained in Figure 6 is similar to the Energy monitoring output in Figure 3. So it is compared to the optimization output, as already performed in Figure 3. The AI is helpful in the comparison and decides the further action. It is also beneficial in the data analytics and forecasting required for the optimization process. The information so obtained is compared with the benchmark. If the benchmark is met, the process stops; else, the process is repeated.

The process is helpful for the management to review the decisions. Also, it helps in analyzing any patterns in the energy management outputs. The different decisions and information can be classified and clustered, which helps analyze the decisions in the future.

Though there are advantages of using AI, the risks of using AI are also involved. The different risks involved are related to Data Security, Data Privacy, Cyber Security, etc. (Abhinav et al., 2021).

Conclusion

This chapter provides a brief overview of the energy management system. The energy management system has evolved over a period of years due to changing electricity industry. New challenges have emerged with the rising importance of data in the electrical system. AI allows utilizing the data and solving the energy management problem in a modern energy management system.

References

Abhinav, J., Sapna, J., Bikram, B., Vishal and J. Hemant, U. (2021). Artificial intelligence and cybersecurity: current trends and future prospects. *The Smart Cyber Ecosystem for Sustainable Development*, 431-441.

Abouzar, E., Edoardo P. and Luca B. (2018). Fault detection, isolation and restoration test platform based on smart grid architecture model using internet-of-things approaches. In *IEEE international conference on environment and electrical engineering and IEEE industrial and commercial power systems Europe (eeeic/i&cpseurope),* 1-5.

Andreas, K. and Michael, H. (2020). Rulers of the world, unite! The challenges and opportunities of artificial intelligence. *Business Horizons*, 37-50.

Armaroli, N. and Vincenzo, B. (2007). The future of energy supply: challenges and opportunities. *Angewandte Chemie International Edition,* 52-66.

Asif, M. and Tariq, M. (2007). Energy supply, its demand, and security issues for developed and emerging economies. *Renewable and sustainable energy reviews*, 1388-1413.

Cheng, C., Yuhan, H., Marimuthu, K., and Pryan, M. K. (2021). Artificial intelligence on economic evaluation of energy efficiency and renewable energy technologies. *Sustainable Energy Technologies and Assessments*, 101358.

Doty, S. and Wayne C. (2004). Turner. *Energy management handbook. CRC Press.*

IEA. (2022). "*International Energy Agency 2022.*" Accessed February 16. https://www.iea.org/news/renewable-electricity-growth-is-accelerating-faster-than-ever-worldwide-supporting-the-emergence-of-the-new-global-energy-economy.

Imaginovation Insider. (2022). "*The role of AI technology in improving the renewable energy sector.*" Accessed March 01. https://imaginovation.net/blog/artificial-intelligence-in-renewable-energy.

Kari, A. and Arto, S. (2006). Distributed energy generation and sustainable development. *Renewable and sustainable energy reviews*, 539-558.

Leal Filho, W., Balogun, A. L., Surroop, D., Salvia, A. L., Narula, K., Li, C., Hunt, J. D., Gatto, A., Sharifi A., Feng, H., Tsani S. and Azadi, H., (2022). Realising the Potential of Renewable Energy as a Tool for Energy Security in Small Island Developing States. *Sustainability*, 14 (9), 4965.

Makala, B. and Tonci B. (2020). *Artificial intelligence in the power sector.* International Finance Corporation, Washington, DC.

Manivannan, P., Prabha, D., and Balasubramanian, K. (2022). Artificial intelligence databases: turn-on big data of the SMBs. *International Journal of Business Information Systems*, 39(1), 1-16.

Ramyar, R. M., Alan F., Farah, M. and Kaamran, R. (2014). A survey on advanced metering infrastructure. *International Journal of Electrical Power & Energy Systems*, 473-484.

Sara, B. M., Mònica, A. P., Íngrid, M. C., Pau, L. G., Eduard, B. M. and Roberto, V. R. (2021). Artificial intelligence techniques for enabling Big Data services in distribution networks: A review. *Renewable and Sustainable Energy Reviews*, 111459.

Sebastian, P. R. and Shafiuzzaman, K. K. (2018). Development of community grid: Review of technical issues and challenges. *IEEE Transactions on Industry Applications*, 1171-1179.

Steven, T. C. (2015). Research agenda into human-intelligence/machine-intelligence Governance. In *Proceedings of the International Annual Conference of the American Society for Engineering Management.* American Society for Engineering Management (ASEM).

Syed, S. and Bong, J. C. (2020), State-of-the-art artificial intelligence techniques for distributed smart grids: A review. *Electronics*, 1030.

World Energy Climate Statics-Yearbook (2021). "*World Energy Climate Statics-Yearbook* (2021)." Accessed February 16. https://yearbook.enerdata.net/electricity/electricity-domestic-consumption-data.html.

World Energy Climate Statics-Yearbook (2021). "*World Energy Climate Statics-Yearbook 2021 Electricity Consumption Rate.*" Accessed February 16. https://yearbook.enerdata.net/electricity/share-electricity-final-consumption.html.

Yaosuo, X., Liuchen, C., and Julian M. (2007). Dispatchable distributed generation network-a new concept to advance DG technologies. *IEEE Power Engineering Society General Meeting*, 1-5.

Yimy, E. G. V., Rodollfo, D. L. and Josel, B. A. (2019). Energy management in microgrids with renewable energy sources: A literature review. *Applied Sciences,* 3854.

Biographical Sketches

Subhash Yadav received his M.Tech degree from Indian Institute of Technology Roorkee. He is currently pursuing his PhD Degree from NIT Kurukshetra. His research interest includes Renewable Energy Systems and Energy Management.

Pradeep Kumar received the B.E degree in electrical engineering from G. B. Pant Engineering College Pauri Garhwal, India, in 2009, and the M.Tech. and PhD degrees from MNNIT Allahabad in 2011 and 2016 respectively. Currently, he is an Assistant Professor in the Department of Electrical Engineering at National Institute of Technology (NIT) Kurukshetra, Haryana, India. His research interests include Energy Management and Power System Monitoring.

Ashwani Kumar received the B.Tech degree in Electrical Engineering from G. B. Pant University, Pant Nagar, India, in 1988, M.Tech. degree in power systems from Punjab University, Chandigarh, India, in 1994 and Ph.D degree from Indian Institute of Technology, Kanpur, India. Currently, he is a Professor in the Department of Electrical Engineering at National Institute of Technology (NIT)-Kurukshetra, Haryana, India. His research interests include power system deregulation optimization and power system dynamics.

Chapter 2

Issues and Challenges of Latest Green Energy Technology Such as Fuel Cell, Waste to Energy and Application of AI

S. P. Singh*, **Arif Iqbal, Yudhishthir Pandey, Mohammed Aslam Husain and Jaswant Singh**
Electrical Engineering Department, Rajkiya Engineering College, Ambedkar Nagar (U.P.), India

Abstract

The ever-growing energy demand for human needs has resulted in an increasing use trend of renewable energies. The use of renewable energies, particularly solar, wind etc., must be optimized with reduced effects on environmental and living creatures. For the last two decades, research has been tremendously conducted to replace all the conventional sources to decrease the dependency on exhaustible fossil fuel and their harmful environmental effects. Hence, green energy technology is considered a future alternative to meet all energy needs and related services. The present and future renewable energy (RE) goals depend on many factors. This includes the technology for optimized energy extraction from natural resources and better management and distribution systems. In light of future renewable energy (RE) goals, artificial intelligence (AI) is now the focus of present research and developments. This chapter deals with the critical issues and challenges of the latest Green Energy Technology with their applications using AI

---

*Corresponding Author's Email: singhsurya12@gmail.com.

In: Applied Artificial Intelligence (AI) to Green Power Technology
Editors: Yogesh Kumar Chauhan, Ranjan Kumar Behera and Asheesh K. Singh
ISBN: 979-8-88697-131-6
© 2022 Nova Science Publishers, Inc.

Keywords: Green energy technology, renewable energy, artificial intelligence (AI), technical challenges

Introduction

Globally, a nation's economy depends directly or indirectly on electric power generated effectively with an appropriate system of management and distribution [1-3]. Although power is developed through the conventional source to meet humankind's energy demand, it results in massive ill effects on the environment and climatic change. Consequently, using clean energy is considered a feasible, user-friendly solution with minimum contamination by CO_2 and hence reduces the greenhouse effect to prevent environmental degradation [4-6]. Consequently, it leads to tremendous research work in green energy at various levels and sublevels of government and the public to achieve enhanced efficiency, guaranteeing future energy demand. Uses of green energy technology are more feasible to implement with a simple, cost-effective system requiring little maintenance and unlimited sources. With time, adaptability is a necessary condition for their additional development. Technology scope is improving daily, and applications in the previous system have become an integral part of the research focus. Presently, the concept of artificial intelligence AI is tremendously used to develop the intelligent machine system with necessary software tools in specific applications [7]. AI plays a vital role in the technical development of renewable energy. This includes predicting energy requirement, heating load, forecasting short load electric power, photovoltaic system sizing, electrical load predictions, etc. This chapter focuses on the uses of green energy technology with the applications of AI, highlighting important issues and challenges in different aspects.

Fuel Cell

Fuel cells are among the most promising clean energy sources, with almost zero carbon emission value. Moreover, as compared to batteries, fuel cells are lighter and smaller, and their refuelling is easier. The high energy density of fuel cells attracts the interest of many researchers; however, the high cost, durability and performance issues hamper its commercialization [8]. The proper selection of materials, likely reversal of voltages and avoiding fuel starvation can lead to better performance of fuel cells. The proton exchange membrane fuel cell (PEMFC) is the most widely used. The sub-systems used in PEMEC-based systems are mainly power electronics and heat and water management. Multiple fuel cells are stacked to obtain the required rating voltage and current. However, the polarization of electrodes reduces the voltage level in case of overloading on the fuel cell stack. This requires the control of the voltage by either controlling the hydrogen or air flow rates.

The control strategies are mainly of two types, the first one is for improving the electrochemical reactions, and the second one is for fixed voltage operation. To improve the electrochemical reactions, humidity control, hydrogen or air flow rate control, avoiding fuel starvation, maintaining proper temperature and anode purging during start-ups are mainly done. Fixed voltage operation is obtained using either supercapacitors or batteries. The main sub-systems of a PEMFC are shown in Figure 1.

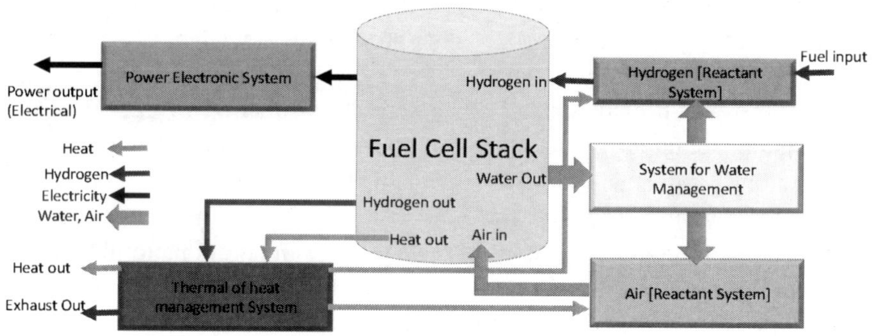

Figure 1. Layout diagram of PEMFC-based system.

One of the main concerns with the PEMFC system is to avoid starvation of the fuel (hydrogen) and oxidant (air/oxygen). There are various methods to look after this concern of avoiding starvation. The control strategies mainly adopted for this are: Classical feedback and feed-forward control using PID control [9-18]; Adaptive control [8, 14, 19-22]; Model predictive control [23-26]; Neural network control [27-30]; Fuzzy logic control [31-46]; Fuzzy logic-PID control [47-51].

Artificial Intelligence Techniques

Artificial intelligence deals with the study of making computers do things intelligently, which, in a movement, people do the better [52-53]. It comprises the expert system, a computer program with information from some experts because of solving the problems and giving advice. An expert system is built using software packages to translate the expertise of humans into the computer programme. Uses of the AI technique in renewable power generation systems have been an active research area for the last few decades, including different methodologies, some of which are discussed in the following sections.

Artificial Neural Networks

It is considered an information processing system inspired by the model's formulation similar to the working of the human brain. It consists of many interconnected neurons network as data processing elements. In ANN, data is passed through different layers via synapsis, characterized by their weight/strength. It also consists of an activation function to limit the output magnitude. Required relation between input and output is obtained from the connection weight and activation functions. This process is known as supervised learning. ANN is not limited to a specific computer task but is trained concerning the available data sets to learn the input pattern. ANN also has an automatic learning capability to recognize the data pattern of natural/physical systems or other sources.

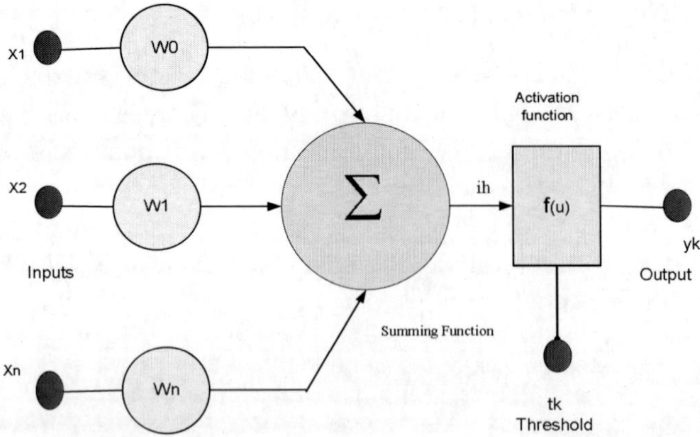

Figure 2. A simple processing element.

Multi-Layer Perceptrons (MLPs)

Multi-Layer Perceptrons (MLPs) are popularly used feed-forward networks. A three-layer MLP is shown in Figure 3. Which consist of input, output and a hidden layer? Input signal obtained in input layer through neurons works as a buffer to distribute the signal to hidden layer through their neurons. In general, any number of the hidden layer (mostly 1 or 2) is possible in ANN. A network's neurons are linked to their next layer through all neurons, as shown in Figure 3. Mathematically one output obtained in terms of neuron x with n inputs is given by

$$y(x) = f(\sum_{i=0}^{n} w_i x_i) \tag{1}$$

where w_0---------------w_n: input weights

f is a nonlinear activation function [1995-1996].

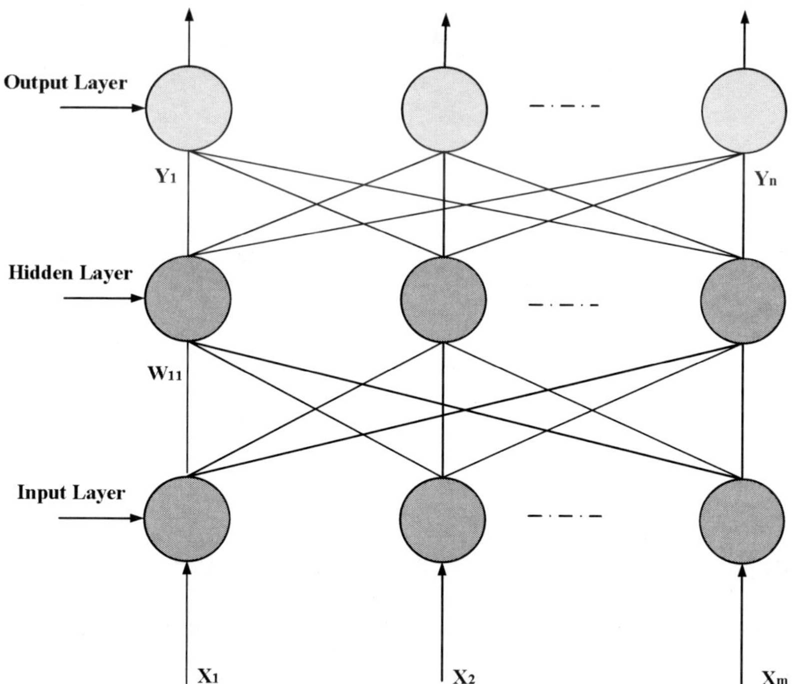

Figure 3. An architecture of Artificial Neural Network (ANN).

The activation function is usually a step or sigmoid function. The commonly sigmoid function is used, which is anti-symmetric about (0, 0.5) and asymptotic about 0 and 1:

$$f(x) = \frac{1}{1+\exp(-\beta x)} \tag{2}$$

In implementing ANN, learning is achieved by using the training algorithm following the learning law by simulating the learning mechanism of the biological system [54-55]. Neural Network is capable of learning for complex tasks like pattern recognition, trend prediction, system identification and process control [54-57]. In this process, neurons of all hidden and output layers are processed by multiplying each input by its weight, and the product is summed up to generate the production through the activation function. The flow of information is unidirectional in feed-forward ANN and in both directions in the case of Feedback ANN [58-59]. During the learning process, a generalized delta rule (G.D.R.) based learning algorithm is used in back-propagation Multi-Layer Perceptrons (MLPs). G.D.R. minimizes the error by the summation of the error square of all output units as defined by

$$E_P = \sum_k \left(y_{pk} - o_{pk}\right)^2 \tag{3}$$

The negative gradient of equation (3) w.r.t. weight W_{kj} determines the direction of weight shifting in the output layer.

Radial Basis Functions (RBF)

The RBF function is suitably used for pattern classification [60]. A three-layer basic architecture of an RBF network is shown in Figure 4. The input pattern (x_1 to x_n) is received by neurons in the input layer. An arbitrary basis is constituted to the input pattern through a set of activation functions by the hidden layer neurons. The distance between the input vector and the centre of each activation or basis function is calculated at the input of each hidden neuron. The output of the hidden neuron is obtained after applying the basis function to this calculated distance. Output y_1 to y_p of the network is obtained from the output layer neurons as a weighted sum of hidden layer activation

[55, 61]. Commonly, the Gaussian distribution function is used as a basis function, and the output of network y_j is mentioned as

$$y_j = \sum_{i=1}^{k} W_{ji} K\left(\frac{\|x - c_i\|}{\sigma_i}\right) \tag{4}$$

W_{ji}: weight of hidden neuron i to output j

σ_i : spread of the function

C_i: centre of the basic function

K(x) = activation function

RBF network is trained to determine the neuron weight W_{ji}, C_i and spread σ_i to obtain the correct output y_j for input pattern x. RBF network training also involves an error function minimization.

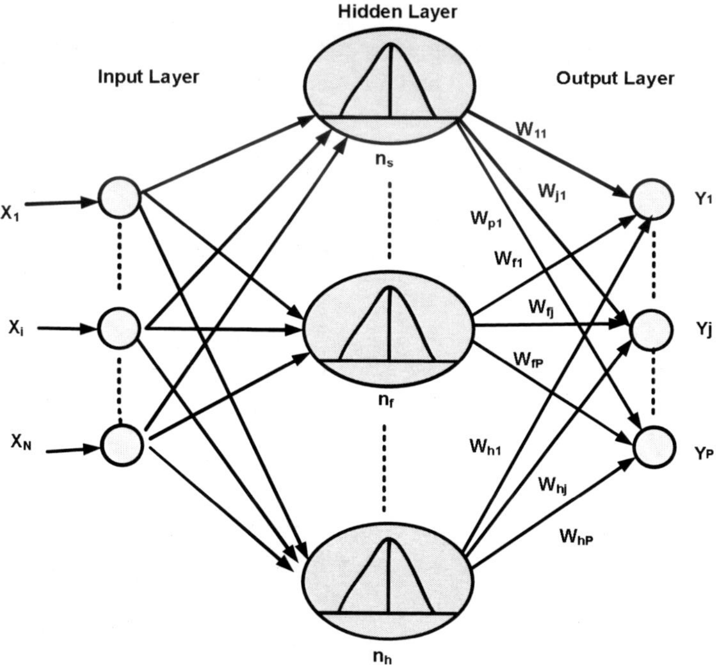

Figure 4. RBF network topology.

Fuzzy Logic

It is a multi-valued logic derived from the theory of fuzzy sets with more approximate reasoning than precise. It is based on linguistic rules using IF-THEN-ELSE statements. It is mainly used in control applications where input parameters or system models are unknown and highly unstable. Understanding and modifying fuzzy controllers' set rules is convenient, which can use the human operator's strategy with natural linguistic terms.

Fuzzy logic offers several attractive features for use in control problems.

a) It constitutes a robust system that does not require precise, noise-free inputs and provides a fail-safe in the non-operation of feedback sensors.
b) User-defined rules are implemented in the F.L. controller, which can be modified to improve or change the system performance.
c) Any sensor which provides the data regarding action and reaction of the system is sufficient, allowing the use of inexpensive sensors resulting in low cost and complexity of an overall system.
d) F.L. provides a controlled operation of a nonlinear system, whose modelling is difficult and tedious.

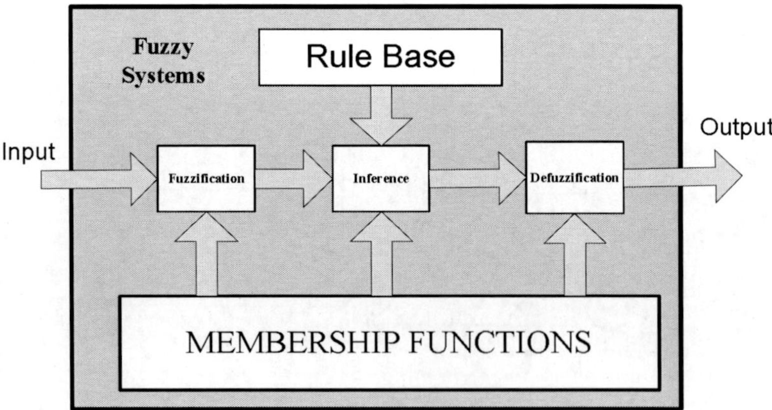

Figure 5. Major components of a fuzzy system.

Fuzzy systems deal with imprecise/incomplete data by using a fuzzy set. Unlike the conventional set theory, an object can take any value between 0 and 1 using fuzzy membership. Major components of a fuzzy system are shown in Figure 5. During fuzzification, exact (crisp) input values are transformed into fuzzy membership [62-63]. Fuzzified data using a Fuzzy interference machine are combined with a prior rule to obtain fuzzy output converted to a crisp value known as defuzzification. Defuzzification can be maximum, mean of maximum and centroid.

A fuzzy logic system can handle imprecise or vague information, which is considered its major strength compared to other AI techniques. However, the significant difficulty of a fuzzy system is determining a good membership function. Also, a fuzzy system possesses no learning capability. To overcome these limitations, modelling of fuzzy systems is combined with other techniques resulting in a hybrid system [63-65].

AI Applications in Renewable Energy

AI in Solar Energy

Although solar radiation is an essential parameter of solar P.V. systems, it has no proper measuring mode. Presently, to measure solar radiation, different climatic parameters are used. For this purpose, ANN [66] is significantly used to predict solar radiations by using the data related to climatological and meteorological obtained from weather station data and satellites. ANNs are mainly of different types to expect solar radiations: back-propagation learning, multi-layer perceptron (MLP) [67], feed-forward algorithm [68], and adaptive model [69]. In the basic ANN model, collected data are inputted to the input layer, and final data is obtained from the output layer to the computer through multiple hidden layers. Building ANN consists of collecting input data, selecting network architecture, and learning algorithms followed by the training test. ANN model used to forecast solar irradiance hourly depends on the time series irradiance data [70]. Three approaches were used to develop a solar irradiance ANN system based on past observational data meteorological parameters [71-77] and both [78-81]. Most developed models forecast hourly, requiring huge meteorological parameters, which are sometimes unavailable [82]. This issue is overcome by using a multi-layer perceptron (MLP) which may forecast the solar irradiance one day (daily) in advance. The best model for predicting solar irradiation having three neurons is given in [83]. The results range lies with a correlation coefficient of 98% and 95% for sunny and cloudy days, respectively. This implemented method highlights the excellent insight of planning and operation of the sustainable renewable system.

Presently, research is focused on developing an online forecasting system for solar P.V. power generation. It generally consists of two-stage approaches

a) At different times solar irradiance is forecasted based on ANN, Fuzzy logic, and hybrid system [84-85]
b) Meteorological and past data are used to develop a regression model

In recent years, most of the developed models are based on the recurrent neural network [86-87] using past data [88] and integrated ANN model with wavelet analysis [89]. Models are also specially designed by using the integration of Marko transition matrix and multi-stage ANN, two-dimensional models [90] and diagonal recurrent wavelet neural network (DRWNN) [91].

Nowadays, forecasting of solar irradiance is done on an hourly basis of present and next day using fuzzy adaptive resonance theory (A.R.T.) for the integration of the Distributed Intelligent Energy management System (DIEMS) [92]. This model can forecast output power up to only level 7, making it less accurate [89]. To overcome this disadvantage, the Radial basis Function Network (RBFN) [89] is used to forecast the generated power for 24 hours. This system indicates the output power using past data and meteorological services. The predicted value and operating data of P.V. devices are compared for efficiency verification. Use of the neural network for prediction purpose of solar P.V. system which is combined with fuzzy rules [93], wavelet transform [94], recurrent W.N.N. [95], firefly algorithm and fuzzy A.R.T. mapped with wavelet transform [96]. Hence, there are many methods for predicting solar P.V. generation having some limitations. The development of a more accurate prediction model remains demanded. This has resulted in introduction of a metaheuristic optimization algorithm, particularly the Shark Smell Organisation (SSO) algorithm [97] was presented for an optimized solar P.V. forecasting by improving the search capability both locally and globally. The proposed metaheuristic algorithm used in the model was found to be most efficient during forecasting of solar P.V. system in comparison with other algorithms [98].

AI in Wind Energy

Condition Monitoring System (C.M.S.) using ANN used for a wind turbine system is given in reference [99]. A regular ANN model usually consists of a single input with 20 neurons and one hidden layer. Prediction technique, mainly Mahalanobis distance, was used during anomaly detection with exogenous information to the nonlinear autoregressive exogenous (NARX) ANN. Hence, continuous monitoring of the turbine using ANN can evade disastrous failure many hours in advance [99]. The genetic algorithm using a multi-layer autoregressive neural network, fuzzy logic, and SIMAP was based on the financial and technical standards [100] and operated according to the maintenance calendar. Out of extensive SCADA data, the only significant, interesting aspect is provided to the operator for easier fault detection [101]. Feed-forward and NARX ANN with multiple hidden layers were used to monitor the wind turbine system [102].

With the ever-increasing energy demand and fear of extending non-renewable sources, needs will be fulfilled by using renewable energy sources. For the accumulation of bulk power, wind speed forecasting is significant using a wind power generation system. Wind speed prediction can be improved by using a unique emplacement technique of wind pattern and electric power [103].

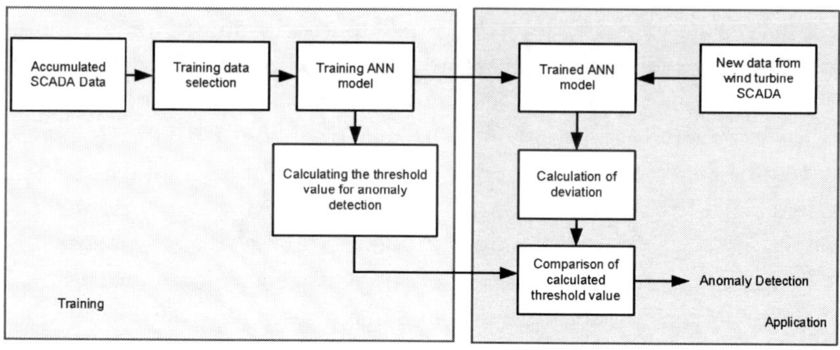

Figure 6. Condition monitoring method of ANN using SCADA data [99].

The MLP training algorithm is used to predict wind speed. Research attention towards the uses of MLP has increased due to their operation accompanied by deep learning. Adjusting spatial weight to the output and input of MLP is well suited for environmental parameters [104]. Feedback and feed-forward ANNs are mainly used to forecast wind speed [105]. For wind speed estimation, the developed ANN model was fed with many parameters, particularly longitude and latitude, elevation, and maximum/ minimum temperature. The accuracy of the obtained results was 95.2% and 93.2% for feed-forward and feedbacked Neural networks, respectively [106]. The Kalman filter tool is also used to estimate the linear system using Gaussian white noise [107]. This recursive tool is made suitable for nonlinear application with proper modifications. The ensemble Kalman filter (EnKF) is used for wind speed prediction. A better result is obtained if this method is combined with ANN. For better accuracy in results during wind speed prediction, the output of ANN is modified using EnKF as a correction structure [108]. Results obtained using ANN to predict momentary wind speed are extrapolated for long-term applications using Markov chains. With the L.M. optimization technique, better wind power prediction using ANN is possible considering different physical parameters like wind speed, direction, pressure, and power [109].

AI in Geothermal

Geothermal energy is obtained through the generated heat inside the earth's crust and atmospheric temperature accompanied by gases, minerals and salts [110]. Geothermal resources at high temperatures are used to produce electricity, whereas, with lower and medium temperatures, they are used as direct fields [111]. With recent research, more insufficient temperature resources are also used with heat pumps. Globally, geothermal power and electricity generation increased to 17% and 10% from 2010 to 2015 [112].

Geothermal power plant efficiency was optimized through the A.B.C. model [112-116]. This model was used in comparison with the conventional system for an advanced analytical method [112]. The presented study highlights a closed agreement of energy efficiency among advanced and developed A.B.C. models. It was found that an arbitrary value is obtained by using an advanced model, whereas a constraint value between the minimum and maximum ranges is obtained [112].

A.B.C. algorithm [117-118] is considered an advanced algorithm to obtain the desired efficiency (Thermal and operational) of the Kalina geothermal cycle [119-120]. The A.B.C. algorithm provides merits in multi-objective, multidimensional function, and multi-model issues compared to other popular optimization methods [121]. Optimized thermal efficiency of 20.36% is achieved using the A.B.C. algorithm [121].

In earth science [122-126], ANN usage is drastically increasing in the geothermal and petroleum domain [126-127]. Static formation Temperature (S.F.T.) of geothermal wells is predicted using three-layer ANN. The ANN training was carried out using the available database of S.F.T. of geothermal borewells [128]. S.F.T. estimation was done by using seven analytical methods [129]. To normalized S.F.T., statistical test [130], the mean value of S.F.T., and deviations in regular estimations are used with statistical rejection process/enhanced out layer identification [131]. The statistical method detects the errors by integrating with ANN [128]. The architecture of Multilayer feed-forward was also used in the ANN model, having more advantages than the analytical method [128]. S.F.T. prediction using the ANN approach results in a faster practical tool in geothermal applications [128].

Challenges in AI Techniques for Green Energy

a) *Availability of data:* The performance of AI-based models, particularly ML during deep learning, is prominently constrained by the quality data availability during training and model evaluation [132]. In data-driven approaches, the availability of quality data, particularly in the newly designed system, offers challenging issues for green energy. Further, the training algorithm uses generalization, and overfitting affects the results.

b) *Data Auditing:* Data analysis for feasibility and quality is critical during predictive maintenance of green energy systems. The data measurement is accompanied by the noise produced from the system's mechanical part. Therefore, multi-variant data should be identified with noise during the measurement analysis. Associated risk due to noise and their performance impact on the predictive model is accessed. For this purpose, frequent audit/test is needed to ensure the algorithm's validity and required learning [133].

c) *Feature Selection:* Selection of a feature is crucial during predictive maintenance required in a renewable energy system. The outcome of the data relating to the critical part or variable is identified. Feature selection was challenging and conducted automatically or manually under expert supervision. The approach adopted for feature selection requires knowledge in the technical domain highlighting fault conditions and methodology, which varies from one model to another. A detailed understanding of technical knowledge is unnecessary for neural networks, whereas a medium account is required in expert and fuzzy systems. A straightforward experience in the knowledge domain is necessary for a physical model.

d) *Multiple faults /failure modelling:* Multiple fault/failure modelling is very complex due to sizeable real-time data sets affecting algorithm performance. A good algorithm and suitable resources may address this challenge.

e) *Robust model design:* Intuition-based results of an unknown system's features are unsuitable, requiring a powerful model of existing green energy resources. It is more challenging to design a robust model of green energy which can handle and manage the fault condition.

f) *Adversarial Attacks:* Adversarial and security attacks are the significant challenges in different AI based models for the application of smart grid and energy management systems [134-135]. Most of the approaches are based on cloud computing, deep learning and Internet of Things (IoT) devices during the predictive maintenance process of the system. These approaches are more prone to the attacks of adversarial and cyber security [136].

g) *Performance and Prediction:* Predictive maintenance of green energy systems is also a significant challenge where ML model prediction should be explainable with a proper sense to a human. However, most of the literature has used the rule-based and traditional ML techniques during predictive maintenance of the system. Few latest works are based on the approach of deep learning. Rule-based methods and conventional algorithms are explainable with lower accuracy during prediction [129, 137-140]. On the other hand, deep knowledge is very effective in prophecy, but the prognosis is not explained. Maintaining equilibrium between performance and decision explanation in predictive maintenance of green energy is difficult.

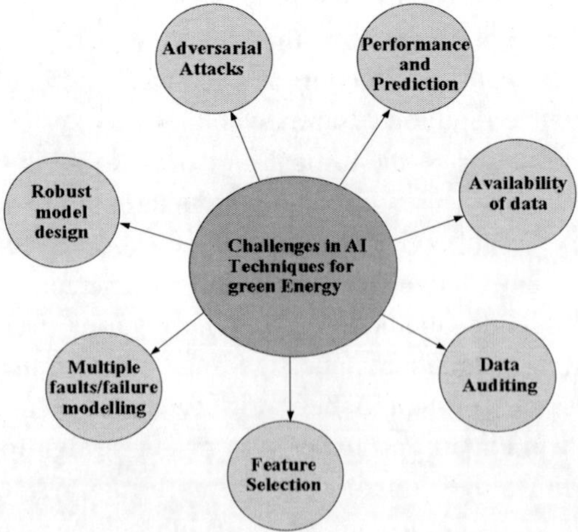

Figure 7. Key challenges and issues in AI based green energy system.

Conclusion

The conventional power industry faces many problems due to specific issues beyond the utility's control. The advancement in Artificial Intelligence (AI) technique will be beneficial for the power industry to optimize the operation of the power grid by reducing its maintenance and increasing its reliability. Artificial Intelligence (AI) can be vital in managing new energy systems. Utilities and government policies using AI to maximize the Artificial Intelligence (AI) extensive range of benefits while reducing its detrimental effects and risks is a crucial aspect for society and the economy. This chapter presents green energy such as fuel cell and AI techniques like ANN and Fuzzy Systems. The application of AI in renewable energy resources such as; solar energy, wind energy and geothermal along with their challenges are presented and discussed.

References

[1] Johansson Thomas B., Henry Kelly, Amulya K. N. Reddy, Robert H. Williams "Renewable fuels and electricity for a growing world economy: defining and achieving the potential," *Energy Studies Review* 4, no. 3, 1992.

[2] Wood Allen J., Bruce F. Wollenberg, Gerald B. Sheblé, "*Power generation, operation, and control*," John Wiley & Sons, 2013.

[3] Sorensen B., "*Renewable Energy: Physics Engineering Environmental Impacts Economics and planning*," 4th edition, Amsterdam: Elsevier, 2021.

[4] Chalka S. G., Miller J. F., "Key Challenges and recent progress in batteries, fuel cells, and hydrogen storage for clean energy systems, " *Journal of Power Sources*, 159: 73-80, 2006.

[5] Kamat P. V., "Meeting the clean energy demand: Nanostructure architectures for solar energy conversion," *Journal Physics Chem C;* 111 (7):2834-60, 2007.

[6] Panwar N. L., Kaushik S. C., Kothari S., "Role of Renewable Energy Sources in Environmental Protection: a review," *Renewable Sustainable Energy Rev*.; 15: 1513-24, 2011.

[7] Turban E., Frenzel, "*Expert Systems and Applied artificial Intelligence*," New Jersey: Prentice Hall, 1992.

[8] Kanth, Utkarsh, Gupta, Anand K., Rai, Abhishek, Mishra, A. K., "Performance study of proton exchange fuel cells for different atmospheric conditions," *IEEE International Conference on Electrical, Computer and Communication Technologies (ICECCT)*. IEEE, 2019.

[9] Claire H., Woo J. B. B., "P.E.M. fuel cell current regulation by fuel feed control," *Chemical Engineering Science*, 62: 957-68, 2007.

[10] Serra M., Aguado J., Ansede X., Riera J., "Controllability analysis of decentralized linear controllers for polymeric fuel cells," *Journal of Power Sources*; 151: 93-102, 2005.

[11] Watkins A. N., Gao W., "An experimental environment for studying hybrid fuel cell system operating characteristics," *North American Power Symposium (NAPS)*:1-6, 2009.

[12] Pukrushpan J. T., Stefanopoulou A. G., Huei P., "Modeling and control for P.E.M. fuel cell stack system," *American Control Conference, Proceedings* 4: 3117-22, 2002.

[13] Pukrushpan J. T., Stefanopoulou A. G., Varigonda S., Pedersen L. M., Ghosh S., Huei P., "Control of natural gas catalytic partial oxidation for hydrogen generation in fuel cell applications," *IEEE Transaction on Control Systems Technology*; 13:3-14, 2005.

[14] Jonathan G. Williams, Guoping Liu, Senchun Chai, Rees D., "Design and Implementation of online Self tuning Control for P.E.M. fuel cells," *The world electric vehicle journal;* 2: 7-17, 2008.

[15] Iqbal, M. T., "Modeling and control of a wind fuel cell hybrid energy system," *Renewable Energy*; 28:223-237, 2003.

[16] Tae-Hoon K., Sang-Hyun K., Wook K., Jong-Hak L., Woojin C., "Development of the novel control algorithm for the small proton exchange Membrane fuel cell stack without external humidification," *Applied Power Electronics Conference and Exposition (APEC), Twenty Fifth Annual IEEE*: 2166-73, 2010.

[17] Qiuli Y., Song-Yul C., Srivastava A. K., Wenzhong G., "Improved Modeling and Control of a P.E.M. fuel cell," *Power System for vehicles, South East Conference, Proceedings of the IEEE*P. 331-6, 2006.

[18] Methekar R. N., Prasad V. R., Gudi, D., "Dynamic analysis and linear control strategies for proton exchange membrane fuel cell using a distributed parameter model," *Journal of Power Sources,* 165:152-70, 2007.

[19] Dalvi A., Guay M., "Control and real-time optimization of an automotive hybrid fuel cell power system," *992 Control Engineering Practice*; 17:924-38, 2009.

[20] Zhang J., Liu G., Yu W., Ouyang M., "Adaptive control of the airflow of a P.E.M. fuel cell system," *Journal of Power Sources*;179:649-59, 2008.

[21] Forrai Alexandru, Hirohito Funato, Yukihiro Yanagita, and Yoshitsugu Kato, "Fuel-cell parameter estimation and diagnostics," *IEEE transactions on energy conversion*20, no. 3 668-675, 2005.

[22] Jing S., Kolmanovsky I., "Load Governor for fuel cell oxygen starvation protection: a robust nonlinear reference governor approach," *American Control Conference Proceedings*, vol. 1, pp.828-33, 2004.

[23] Danzer M. A., Wittmann S. J., Hofer E. P., "Prevention of fuel cell starvation by model predictive control of pressure, excess ratio, and current," *Journal of Power Sources*; 190: 86-91, 2009.

[24] Arce A., Ramirez D. R., del Real A. J., Bordons C., "Constrained explicit predictive control strategies for P.E.M. fuel cell systems, Decision and Control," *46th IEEE Conference*, pp. 6088-93, 2007.

[25] Bordons C., Arce A., del Real A.J., "Constrained predictive control strategies for P.E.M. fuel cells," *American Control Conference*, 2006.

[26] Kenarangui Y., Wang S., Fahimi B., "On the impact of fuel cell system response on Power Electronics converter design," *Vehicle Power and Propulsion Conference, VPPC, IEEE,* pp. 1-6, 2006.

[27] El-Sharkh M. Y., Rahman A., Alam M. S., "Neural Networks-based control of active and reactive power of a stand-alone P.E.M. fuel cellpower plant," *Journal of Power Sources*, 135:88-94, 2004.

[28] Caponetto R., Fortuna L., Rizzo A., "Neural Network modelling of fuel cell systems for vehicles," *10th IEEE Conference on Emerging Technology and factory Automations (ETFA)* pp. 187-92, 2005.

[29] Almeida P.E.M., Simoes M. G., "Neural optimal control of P.E.M. fuel cells with parametric CMAC network," *Industry Applications, IEEE Transaction*; 41: 237-45, 2005.

[30] Hatti M., Tioursi M., "Dynamic neural Network controller model of P.E.M. fuel cell system," *International Journal of Hydrogen Energy*; 34: 5015-21, 2009.

[31] Sakhare A., Davari A., Feliachi A., "Fuzzy Logic control of fuel cell for stand-alone and grid connection," *Journal of Power Sources*; 135: 165-76, 2004.

[32] Schumacher J. O., Gemmar P., Denne M., Zedda M., Stueber M., "Control of miniature proton exchange membrane fuel cells based on fuzzy logic," *Journal of Power Sources*; 129: 143-51, 2004.

[33] Jeong K. S., Lee W. Y., Kim C. S., "Energy management strategies of a fuel cell/battery hybrid system using fuzzy logics," *Journal of Power Sources*, 145: 319-26, 2005.

[34] Thomya A., Khunatorn Y., "Design of control System of Hydrogen and oxygen flow rate for Proton Exchange Membrane fuel cell using Fuzzy Logic Controller," *Energy Procedia*, 9:186-97, 2011.

[35] Hajizadeh A., Golkar M. A., "Intelligent Power management strategy of hybrid distributed generations system," *International Journal of Electrical Power & Energy Systems*; 29:783-95, 2007.

[36] Erdinc O., Vural B., Uzunoglu M., Ates Y., "Modeling and Analysis of an FC/UC hybrid vehicular power system using a wavelet-fuzzy logic based load sharing and control algorithm," *International Journal of Hydrogen Energy*, 34: 5223-33, 2009.

[37] Kisacikoglu M. C., Uzunoglu M., Alam M. S., "Load sharing using fuzzy logic control in a fuel cell/ultracapacitor hybrid vehicle," *International Journal of Hydrogen Energy*, 34: 1497-507, 2009.

[38] Li C. Y., Liu G. P., "Optimal fuzzy power control and management of fuel cell/battery hybrid vehicles," *Journal of Power Sources*, 192:525-33, 2009.

[39] Schouten N. J., Salman M. A., Kheir N. A., "Fuzzy logic control for parallel hybrid vehicles," *Control Systems Technology, IEEE Transaction*; 10:460-8, 2002.

[40] Tekin M., Hissel D., Pera M. C., Kauffmann J. M., "Energy Management Strategy for embedded fuel cell system using fuzzy logic," *Industrial Electronics, IEEE International Symposium* p. 501-6, vol.1, 2004.

[41] Zadeh L. A., "Fuzzy Sets," *Inform Control*. 1965, 8:338-53, 1965.

[42] Mo Z., Zhu X., Cao G., "Design and Simulation of fuzzy controller for PEMFCs," Industrial Technology, *ICIT IEEE International Conference*, pp. 220-4, 2005.

[43] Jurado F., Valverde M., "Novel Inverter flux control for fuel cell using Fuzzy logic, Power Electronics and Motion Control Conference," 2004 IPEMC, *The 4th International* 2004, p. 714-8, vol.2, 2004.

[44] Jurado F., Valverde M., "Fuzzy logic inverter flux control of fuel cell plants in distributed generation," *Harmonics and Quality of Power, 11th International Conference*, p.35-40, 2004.

[45] Jonathan G. Williams, Guoping Liu, Senchun Chai, Rees D., "Intelligent Control for Improvements in P.E.M. Fuel cell flow Performance," *International Journal of Automation and Computing,* 2:145-54, 2008.

[46] Mestan T., Daniel H., Marie-Ccile P., Jean Marie K., "Energy-Management Strategy for Embedded Fuel-cell Systems using Fuzzy Logic," *Industrial Electronics, IEEE Transactions* 54: 595-603, 2007.

[47] Agbossou K., Bilodeau A., Documbia M. L., "Development of a control method for a renewable energy system with fuel cell," *AFRICON*. p.1-5, 2009.

[48] Thanapalan, K. K. T., G. C. Premier., A. J. Guwy, "Model Based Controller Design for Hydrogen Fuel-cell Systems," *Renewable Energy and Power Quality Journal.* vol.1, No.9, May 2011.

[49] Zhan Y., Zhu J., Guo Y., Jin J., "Control of Proton Exchange Membrane fuel cell based on Fuzzylogic," *Control Conference*, CCC, 2007 Chinese p.345-9, 2007.

[50] Yuedong Z., Jianguo Z., Youguang G., Hua W., "Performance analysis and improvement of a proton exchange membrane fuel cell using comprehensive intelligent control," *Electrical Machines and Systems, 2008 ICEMS International Conference*, p. 2378-83, 2008.

[51] Rgab, O., Yu, D., and Gomm, J., "Polymer electrolyte membrane fuel cell control with feed-forward and feedback Strategy," *International Journal of Engineering, Science and Technology*, 2:56-66, 2010.

[52] Singh G. K., "Modelling and Experimental analysis of a self-excited six-phase induction generator for stand-alone renewable Energy generation," *Renewable Energy*; 33(7):1605-21, 2008.

[53] Lei M., Shiyan L., Chuanwen J., Hongling L., Yan Z., "A review on the forecasting of wind speed and generated power," *Renewable Sustainable Energy Review*; 13(4): 915-20, 2009.

[54] Belu, R. G., Paun, A., and Belu, A. C., "Neural networks in measurement and instrumentation," (Part I), *Romanian Journal of Information Science and Technology*, 6 (1-2), 61-86, 2003.

[55] Chen, S. H., Jakeman, A., Norton, J. P., "Artificial intelligence techniques: An introduction to their use for modelling environment systems," *Mathematics and Computers in Simulation*, 78, 379-400. https://doi.org/10.1016/j.matcom.2008.01.028.

[56] Kalogirou, S. A., "Artificial Neural networks in renewable energy system applications: A review," *Renewable & Sustainable Energy Reviews*, 5(3), 373-401-401. https://doi.org/10.1016/S1364-0321(01)00006-5, 2001.

[57] Kalogirou, S. A., "*Artificial Intelligence in energy and renewable energy systems,*" Nova Science Publisher, 2007.

[58] Fuller, R. "*Introduction to neuro-fuzzy Systems,*" Heidelberg, Germany: Physica-Verlag, 2000.

[59] Hontoria, L., Riesco, J., Zufria, P., Aguilera, J., "Application of neural networks in the solar radiation field," *16th European photovoltaic for Chemical for Chemical Engineers Elsevier* (Vol. 3, pp.385-408), 2001.

[60] Mohandes, M., Balghonaim, A., Kassas, M., Rehman S., Halawani, T. O., "Use of radial basis functions for estimating monthly mean daily solar radian," *Solar Energy*, 68(2), 161-168.doi: 10.1016/S0038-092X(99)00071-7.

[61] Elizongo, D., Hoongenboom, G., Mcclendon, R. W., "Development of a neural network model to predict daily solar radiation," *Agricultural and Forest Meteorology*, 71(1-2), 115-132.doi: 10.1016/0168-1923(94)90103-1.

[62] Zadeh, L. A., Fuzzy sets. *Information and Control*, 8, 338-353, https://doi.org/10.1016/S0019-9958(65)90241-X.

[63] Robert, F., "*Neural Fuzzy Systems,*" Abo Akademi University, 1995.

[64] Lakhmi, C. J., Martin, N. M., "*Fusion of neural networks, Fuzzy systems and genetic algorithms: Industrial applications,*" C.R.C. Press, L.L.C., 1998.

[65] Tefler, H. H., Kadambe, B. A. S., "Neural network adaptive wavelets for signal processing and classification," *Optimization and Engineering*, 31(9), 1907-1916, 1992.

[66] Yadav, A. K., Malik, H., Chandel, S. S., "Application of rapid miner in ANN based prediction of solar radiation for assessment of solar energy resource potential of 76 sites in North western India," *Renewable Sustainable Energy Rev.*, 52, 1093-1106, 2015.

[67] Azadeh, A., Maghsoudi, A., Sohrabkhani, S., "Using an integrated artificial neural networks model for predicting global radiation: the case study of Iran." *Energy Convers, Manage,* 50, 6, 1497-1505, 2009.

[68] Hasni, A., Abdelkrim, S., Belkacem, D., Bassou, A., "Estimating global solar radiation using artificial neural network and climate data in the south-western region of Algeria," *Energy Proceedings*, 18, 531-537, 2012.

[69] Mellit, A., Eleuch, H., Benghanem, M., Elaoun, C., Massi Pavan, A., "An adaptive model for predicting of global, direct and diffuse hourly solar irradiance," *Energy Conversion Manage*, 51, 771-782, 2010.

[70] Muneer, T., Younes, S., Munawwar, S., "Discourse on solar radiation modelling," *Renewable Sustainable Energy Rev.*, 11, 4, 551-602, 2007.

[71] Sen, Z. K., "*Solar Energy Fundamentals and Modeling Techniques,*" Springer-Verlag, London, 2008.

[72] Husain, M. Aslam, Tariq, A., Hameed, S., Saad Bin Arif, M., Jain, A., "Comparative assessment of maximum power point tracking procedures for photovoltaic systems," *Green Energy & Environment* 2.1 :5-17, 2017.

[73] Ajoy K. Palit, Dobrivoje Popovic, "Computational Intelligence in Time Series Forecasting: Theory and Engineering Applications," In *Advances in Industrial Control*, Springer, 2005.

[74] Husain, M. Aslam, Jain, A., Tariq, A., Iqbal, A., "Fast and precise global maximum power point tracking techniques for photovoltaic system," *I.E.T. Renewable Power Generation* 13.14: 2569-2579, 2019.

[75] Mohandes, M., Rehman, S., Halawani, T. O., "Estimation of global solar radiation using artificial neural networks," *Renewable Energy*, 14, 1-4, 179-184, 1998.

[76] Naseem, M., Husain, M. Aslam, Faiz Minai, A., Neyaz Khan, A., "Assessment of meta-heuristic and classical methods for GMPPT of P.V. system," *Transactions on Electrical and Electronic Materials* 22.3: 217-234, 2021.

[77] Al-Alawi, S. M., Al-Hinai, H. A., "An ANN-based approach for predicting global radiation in locations with no direct measurement instrumentation," *Renewable Energy*, 14, 1-4, 199-204, 1998.

[78] Hontoria, L., Aguilera, J., Zufiria, P., "Generation of hourly irradiation synthetic series using the neural network multi-layer perceptron," *Sol. Energy*, 72, 5, 441-446, 2002.

[79] Hontoria, L., Aguilera, J., Zufiria, P., "An application of the multi-layer perceptron: Solar radiation maps in Spain," *Sol. Energy*., 79, 5, 523-530, 2005.

[80] Mellit, A., Kalogirou, S. A., Shaari, S., Salhi, H., Hadj Arab, A., "Methodology for predicting sequences of mean monthly clearness index and daily solar radiation data in remote areas: Application for sizing a stand-alone P.V. system," *Renewable Energy*, 33, 7, 1570-1590, 2008.

[81] Mellit, A., "Artificial Intelligence Technique for modelling and forecasting of solar radiation data: a review," *Int. J. Artif. Intelligence Soft Computing*, 1, 1, 52, 2008.

[82] Mellit, A., Kalogirou, S. A., "Artificial intelligence techniques for photovoltaic applications: A Review," *Prog. Energy Combust. Sci*., 34, 5, 574-632, 2008.

[83] Fernando, D. A. K. and Jayawardena, A. W., "Runoff forecasting using RBF Networks with O.L.S. Algorithm," *J. Hydrol. Eng*., 3, 3, 203-209, 1998.

[84] Bacher, P., Madsen, H., Nielsen, H. A., "Online short-term solar power forecasting," *Sol. Energy*, 83, 10, 1772-1783, 2009.

[85] Mellit, A., Benghanem, M., Arab, A.H., Guessoum, A., "A simplified model for generating sequences of global solar radiation data for isolated sites: using artificial neural network and a library of Markov transition matrices approach," *Sol. Energy*, 79, 5, 469-482, 2005.

[86] Yona, A., Senjyu, T., Funabashi, T., "Application of recurrent neural network to short-term-ahead generating power forecasting for photovoltaic system," *IEEE Power Engineering Society General Meeting*, pp.3659-3664, 2007.

[87] Cao, J. C., Cao, S. H., "Study of forecasting solar irradiance using neural networks with pre-processing sample data by wavelet analysis," *Energy,* 31, 15, 3435-3445, 2006.

[88] Mellit, A. and Pavan, A. M., "A 24-h forecast of solar irradiance using artificial neural network: Application for the performance prediction of a grid-connected P.V. plant at Trieste, Italy," *Sol. Energy*, 84, 5, 807-821, 2010.

[89] Chen, C., Duan, S., Cai, T., Liu, B., "Online 24-h solar power forecasting based on weather type classification using artificial neural network," *Sol. Energy*, 85, 11, 2856-2870, 2011.

[90] Hocaoglu, F. O., Gerek, O. N., Kurban, M., "Hourly solar radiation forecasting using optimal coefficient 2-D linear filters and feed-forward neural networks," *Sol. Energy*, 82, 8, 714-726, 2008.

[91] Cao, J., Lin, X., "Application of the diagonal recurrent wavelet neural network to solar irradiation forecast assisted with fuzzy technique," *Eng. Appl. Artificial Intelligence*, 21, 8, 1255-1263, 2008.

[92] Charaborty, S., Weiss, M. D., Simoes, M. G., "Distributed Intelligent Energy Management System for a Single-phase High-frequency AC Micro-grid," *IEEE Transactions of Industrial Electronics,* 54, 1, 97-109, 2007.

[93] Chaabene, M., Ammar, M. B., "Neuro-fuzzy dynamic model with Kalman filter to forecast irradiance and temperature for solar energy systems," *Renewable Energy*, 33, 7, 1435-1443, 2008.

[94] Mandal, P., Madhira, S. T. S., Haque, A. U., Meng, J., Pineda, R. L., "Forecasting power output of solar Photovoltaic system using wavelet trasnform and artificial intelligence techniques," *Proceedings of Computational Science*, 12, 332-337, 2012.

[95] Cao, J., and Lin, X., "Study of hourly and daily solar irradiation forecast using diagonal recurrent wavelet neural networks," *Energy Conversion Manage.*, 49, 6, 1396-1406, 2008.

[96] Haque, A. U., Nehrir, M. H., Mandal, P., "Solar P.V. power generation forecast using a hybrid intelligent approach," *Power and Energy Society General Meeting (P.E.S.), Vancouver, BC*, 2013.

[97] Abedinia, O., Amjady, N., Ghasemi, A., "A new meta-heuristic Algorithm based on shark smell optimization," *Complex J.*, 21, 5, 97-116, 2016.

[98] Abedina, O., Amajady, N., Ghadimi, N., "Solar energy forecasting based on hybrid neural network and improved metaheuristic algorithm," *Computational Intelligence*, 34, 1, 242-260, doi.:10.1111/coin.12145, 2017.

[99] Bangalore, P., Letzgus, S., Karlsson, D., Patriksson, M., "An artificial neural network-based condition monitoring method for wind turbines, with application to the monitoring of the gearbox," *Wind Energy*, 20, 8, 1421-1438, 2017.

[100] Garcia, M. C., and Sanz-Bobi, M. A., Del Pico, J., "SIMAP: Intelligent System for Predictive Maintenance," *Comput. Ind.*, 57, 6, 552-568, 2006.

[101] Xaher, A., McArthur, S. D. J., Infield, D. G., Patel, Y., "Online wind turbine fault detection through automated SCADA data analysis," *Wind Energy,* 12, 6, 574-593, doi: 10.1002/we.319, 2009.

[102] Karlsson, D., "*Wind turbine performance monitoring using artificial neural networks with a multidimensional data filtering approach*", Masters' Thesis, Dept. of Energy and Environment, Chalmers University of Technology, Gothenburg, Sweden, 2015.

[103] Zhu, X., Genton, M. G., Gu, Y., Xie, L., "Space-time wind speed forecasting for improved power system despatch," *TEST*, 23, 1, 1-25, 2014.

[104] Govorov, M., Beconyte, G., Gienko, G., Putrenko, V., "Spatially contrained regionalization with multi-layer perceptron," *Trans., G.I.S.*, 23, 5048-1077, 2019.

[105] Welch, R. L., Ruffing, S. M., Venayagamoorthy, G. K., "Comparison of feed-forward and feedback neural network architectures for short term wind speed prediction," *International Joint Conference on Neural Networks*, 2009.

[106] El Shahat, A., Haddad, R. J., Kalani, Y., "An artificial Neural Networks model for wind energy estimation," *Southeast Conference*, 2015.

[107] Li, Q., Li, R., Ji, K., Dai, W., "Kalman Filter and its applications," *8th International Conference on Intelligent Networks and Intelligent Systems (ICINIS),* 2015.

[108] Sharma, D. and Lie, T. T., "Wind Speed forecasting using hybrid ANN-Kalman filter techniques," *10th International Power & Energy Conference (IPEC),* 2012.

[109] Jain, V., Singh, A., Chauhan, V., Pandey, A., "Analytical study of wind power prediction system by using Feed Forward Neural Network," *International Conference on Computation of Power, Energy Information and Communication (ICCPEIC),* 2016.

[110] Khublaryan, M. G., "Types and properties of waters." *Encyclopedia of life support systems,* vol. 1, p.1e159, EOLSS Publishers Co. Ltd., Oxford, United Kingdom, 2009.

[111] Bertani, R., "Geothermal Power generation in the world 2010e2014 update report," *Geothermics*, 60, 31e43, 2016.

[112] Ozkaraca, O., Kecebas, A., Demircan, C., "Comparative thermodynamic evaluation of a geothermal power plant by using advanced exergy and artificial bee colony methods," *Energy,* 156, 169-180, doi: 10.1016/j.energy.2018.05.095, 2018.

[113] Sadeghi, S., Saffari, H., Bahadormanesh, N., "Optimization of a modified double turbine Kalina cycle by using Artificial Bee Colony algorithm," *Applied Therm. Energy*., 91, 19e32, 2015.

[114] Saffari, H., Sadeghi, S., Khoshzat, M., Mehregan, P., "Thermodynamic analysis and optimization of a geothermal Kalina cycle system using Artificial Bee Colony algorithm," *Renewable Energy*, 89, 154e67, 2016.

[115] Karaboga, D., "*An idea based on honey bee swarm for numerical optimization*," Technical Report-TR06, Erciyes University, Engineering, Kayseri, Turkey, 2005.

[116] Karaboga, D., Akay, B., "A comparative study of artificial bee colony algorithm," *Appl. Math, Comput.,* 214, 1, 108e32, 2009.

[117] Karaboga, D., Basturk, B., "A powerful and efficient algorithm for numerical function optimization: Artificial bee colony (A.B.C.) algorithm," *J. Global Optim.*, 39, 3, 459e471, 2007.

[118] Karaboga, D., Akay, B., "A comparative study of artificial bee colony algiorithm," *Appl. Math. Comput.*, 214, 1, 108e132, 2009.

[119] Zhang, N., Lior, N., "Methodology for thermal design of novel combined refrigeration/Power binary fluid systems," *Int. J. Refrig*., 30, 6, 1072e1085, 2007.

[120] Tillner-Roth, R., Friend, D.G., "A Helmholtz free energy formation of the thermodynamic properties of the mixture (water ammonia)," *J. Phys. Chem. Ref. Data*, 27, 1, 63e96, 1998.

[121] Saffari, H., Sadeghi, S., Khoshzat, M., Mehregan, P., "Thermodynamic analysis and optimization of a geothermal Kalina cycle system using Artificial Bee Colony Algorithm," *Renewable Energy*, 89, 154-167, 2016, https://doi.org/10.1016/j.renene.2015.11.087.

[122] Goutorbe, B., Lucazeau, F., Bonneville, A., "Using neural networks to predict thermal conductivity from geophysical well logs," *Geophys. J. Int.*, 166, 115-125, 2006.

[123] Hsieh, B. Z., Wang, C. W., Lin, Z. S., "Estimation of formation strength index of aquifer from neural networks," *Comput., Geosci.*, 35, 1933-1939, 2009.

[124] Leite, E. P., de Souza Filho, C. R., "Probabilistic neural networks applied to mineral potential mapping for platinum group elements in the Serra Leste region, Caraja's mineral province, Brazil," *Comput. Geosci.*, 35, 675-687, 2009a.

[125] Leite, E. P., de Souza Filho, C. R., "TEXTNN-a MATLAB program for textural classification using neural networks," *Comput. Geosci*, 35, 2084-2094, 2009b.

[126] Morton, J. C., "Boosting a fast neural network for supervised land cover classification," *Comput. Geosci.*, 35, 1280-1295, 2009.

[127] Serpen, Genco, Yildiray Palabiyik, Umran Serpen. "An artificial neural network model for Na/K geothermometer," *Proceedings of the 34th Workshop on Geothermal Reservoir Engineering,* Stanford University, Stanford, U.S.A. 2009.

[128] Bassam, A., Santoyo, E., Andaverde, J., Hernandez, J.A., Espinoza-Ojeda, O. M., "Estimation of Static formation temperatures in geothermal wells by using an artificial neural network approach," *Comput Geosci*, 36, 9, 1191-1199, 2010.

[129] Hiran, Kamal Kant, Deepak Khazanchi, Ajay Kumar Vyas, Sanjeevi Kumar Padmanaban, eds. "*Machine Learning for Sustainable Development*," Vol. 9. Walter de Gruyter GmbH & Co KG, 2021.

[130] Andaverde, J., Verma, S. P., Santoyo, E. "Uncertainty estimates of static formation temperature in borehole and evaluation of regression models," *Geophys. J. Int.*, 160, 1112-1122, 2005.

[131] Verma, Surendra P., Lorena Díaz-González, Rosalinda González-Ramírez, "Relative Efficiency of Single-Outlier Discordancy Tests for Processing Geochemical Data on Reference Materials and Application to Instrumental Calibrations by a Weighted Least-Squares Linear Regression Model," *Geostandards and Geoanalytical Research* 33.1 (2009): 29-49.

[132] Ahmad, Kashif, Maabreh, M., Ghaly, M., Khan, K. "Developing future human-centered smart cities: Critical analysis of smart city security, interpretability, and ethical challenges," arXiv preprint arXiv: 2012.09110(2020).

[133] *Why artificial intelligence is a game-changer for renewable energy*. https://www.ey.com/enps/power-utilities/why-artificial-intelligence-is-a-game-changer-for-renewable-energy. Accessed: 2021-04-13.

[134] Iman Niazazari Hanif Livani, "Attack on grid event cause analysis: An adversarial machine learning approach," *IEEE Power & Energy Society Innovative Smart Grid Technologies Conference (ISGT)*, pages 1–5. 2020.

[135] Gautam Raj Mode, Khaza Anuarul Hoque, "*Adversarial examples in deep learning for multivariate time series regression*," arXiv preprint arXiv:2009.11911, 2020.

[136] Gautam Raj Mode, Khaza Anuarul Hoque, "*Crafting adversarial examples for deep learning based prognostics (extended version)*." *arXiv preprint* arXiv: 2009.10149, 2020.

[137] Balamurugan, S., Ajay Kumar Vyas, Kamal Kant Hiran, Harsh S. Dhiman, eds." *Artificial Intelligence for Renewable Energy Systems*," John Wiley & Sons, 2022.

[138] Arif Iqbal, Girish Kumar Singh, "PSO Based Controlled Six-phase Grid Connected Induction Generator for Wind Energy Generation," *C.E.S. Transactions on Electrical Machines and Systems,* vol. 5, no. 1, pp. 41-49, March 2021.

[139] Jaswant Singh, Surya Prakash Singh, K. S. Verma, Arif Iqbal, Bhavnesh Kumar, "Recent Control Techniques and Management of AC Micro-grids: A critical Review on issues, Strategies, and future trends," *International Transaction on Electrical Energy Systems*, Impact factor 2.86, pp. 1-39, Wiley, July, 2021. https://doi.org/10.1002/2050-7038.13035.

[140] Pelland, S., Remund, J., Kleissl, J., Oozeki, T., Brabandere, K. D., "*Photovoltaic and Solar Forecasting: State of the Art, IEA PVPS Task 14, Subtask 3.1, Report IEA-PVPS T14-01*," The International Energy Agency (IEA), U.S.A., October 2013.

Chapter 3

Voltage Improvement of Short Shunt Self-Excited Induction Generators Using Gravitational Search Algorithms and Genetic Algorithms

Swati Paliwal*,[1] Sanjay Kumar Sinha[1]
and Yogesh Kumar Chauhan[2]

[1]Department of Electrical and Electronics Engineering, Amity University, Noida, Uttar Pradesh, India

[2]Department of Electrical Engineering, Kamla Nehru Institute of Technology, Sultanpur, Uttar Pradesh, India

Abstract

In this chapter, the voltage performance of the induction generator which is used in micro/pico hydropower generation has been improved. The improvement of voltage depends on an optimized value of capacitance, which is required for excitation. The optimized shunt and series capacitance has been identified using two artificial intelligence techniques named Gravitational search algorithm and the Genetic algorithm. The research has been done on the MATLAB environment for the objective function of root mean square error of voltage regulation in the case of resistive and inductive loads. The simulated results validate that the gravitational search algorithm provides better voltage performance as compared to the genetic algorithm.

---

*Corresponding Author's Email: swatipaliwal03@gmail.com.

In: Applied Artificial Intelligence (AI) to Green Power Technology
Editors: Yogesh Kumar Chauhan, Ranjan Kumar Behera and Asheesh K. Singh
ISBN: 979-8-88697-131-6
© 2022 Nova Science Publishers, Inc.

Keywords: Newton Raphson, gravitational search algorithm, genetic algorithm

Introduction

At the world level, the usage of renewable energy has increased because of rising energy demand and the desire to safeguard the environment. Among other renewable energy sources, micro/pico hydro generation has shown consistent progress in decentralized areas.

Hydropower is clean and contributes 19% of world power. The residual water is used for irrigation and other purposes once the electricity is generated. Especially for offshore areas, pico/mini-hydro energy conversion became a cost-effective and suitable choice. It also has a low environmental impact; hence it is safe as compared to biomass energy and the ocean energy. The generation from hydropower has evolved different types of machinery such as Permanent Magnet Synchronous Generator (PMSG), Doubly-fed Induction Generator (DFIG), and the Self-Excited Induction Generator (SEIG).

In remote locations, Induction Machines (IM) have shown numerous advantages. From the principle of Faraday's law, IM obtains its power at the stator terminal. IM can be used in either driving or generating zones depending on slip characteristics. In the motoring region, slip varies as $0< s < 1$, and in generating region $-1< s < 0$. Basset and Potter accounted for the usage of induction generators using external capacitors.

Initially, synchronous generators were used in distant places to convert hydro energy into electrical energy [1]. But for isolated areas, SEIG proved a suitable choice. SEIG has been used as a prime mover to provide an appropriate terminal voltage. SEIG has several advantages like low unit costs and low maintenance. But SEIG also faces disadvantages like high voltage dips, inrush currents, and poor power factors due to the load non-linearity [2]. This imbalanced operation can result in overheating, insulation stress, winding stress, and shaft vibration since phase currents and voltages are not equal. As a result, SEIG's global acceptance is contingent on solutions to address its flaws and its capacity to handle dynamic loads under unbalanced conditions [3]. In this chapter, three phases, 5kW IM operated as SEIG & its voltage has been improved using Gravitational Search Algorithm (GSA) and Genetic Algorithm (GA).

Literature Review

The terminal voltage of SEIG is governed by the capacitor bank used for excitation, the speed of the prime mover, and the connected load. Based on various criteria, capacitance has been chosen for SEIG operation under different loading conditions [4]. But SEIG terminal voltage drops with the change in load because of the growing difference between VAR supplied and VAR generated.

Several voltage-regulating mechanisms have been employed in SEIG, including switching capacitors, variable inductors, saturation reactors, and other semi-conductor-based devices [5]. The voltage regulator regulates the circuit and other difficulties such as harmonics and switching transients, making it inefficient to use [6]. As a result, short and long shunt SEIG configurations have been used to minimize the voltage regulator's complexity [7]. The purpose of series and shunt capacitance in SEIG is to provide regulated and rated voltage at no load [8]. Series capacitors in short-shunt and long-shunt configurations were employed to generate the regulated SEIG voltage.

The positioning of the series capacitor in the SEIG machine determines the configurations. Li-Wang and Ching-Huei Lee [9] investigated the dynamic response of short and long shunt type SEIG feeding with an induction motor load. In order to improve the V/F ratio, steady-state modeling of SEIG has been done [10]. A new optimized meta-heuristic approach has been developed for determining the optimal capacitance value at rated levels. The magnetizing reactance and frequency of SEIG were identified using iterative techniques in this scheme, resulting in a correlation between the computational algorithm and the experimental data. To maximize the value of magnetizing reactance and frequency, Hannan et al., and D. Bouhadjra et al., in paper [11] employed three optimization techniques: Genetic algorithm (GA), Particle Swarm Optimization (PSO), and Simulated Annealing (SA). The resistive load is fed by SEIG, and the optimization is done there in paper [11]. Also, in [12] GA was applied to increase SEIG's performance. In a paper [13] gravitational search algorithm (GSA) with PV excitation management techniques was used in a wind turbine induction generator. S. Paliwal et al., in [14] on the other hand, used GSA to improve SEIG frequency.

SEIG performance becomes sluggish because the frequency and magnetizing reactance change with the load even at constant rotor speed. Therefore, three constraints are necessary for SEIG machine modeling (Equality, Inequality, and Bound Limits). Also, steady-state analysis is essential to find the required parameters to increase its performance. To model equivalent circuits and computation of objective functions, it is important to understand the mathematical modeling of SEIG which is being discussed in the Problem Structure section.

Problem Structure

Figure 1 illustrates a short-shunt SEIG machine. In this figure, capacitance is connected in series to give needed voltage regulation under full load and is connected in parallel to offer no-load voltage. Here, SEIG is connected with the R and R-L type of load.

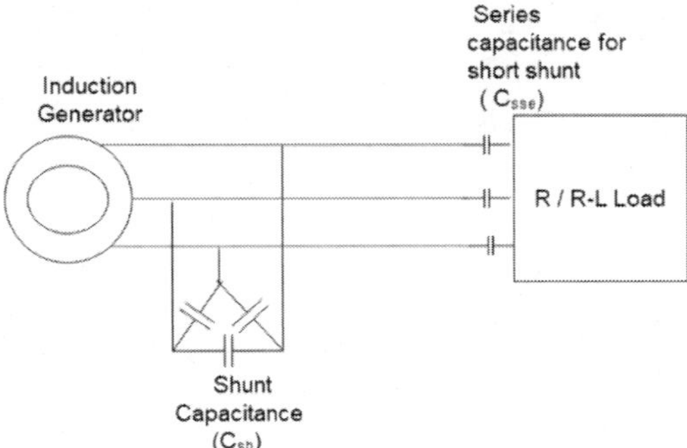

Figure 1. Short shunt SEIG machine.

As shown in Figure 2, the modeling of SEIG was done with the help of the mathematical modelling of the machine.

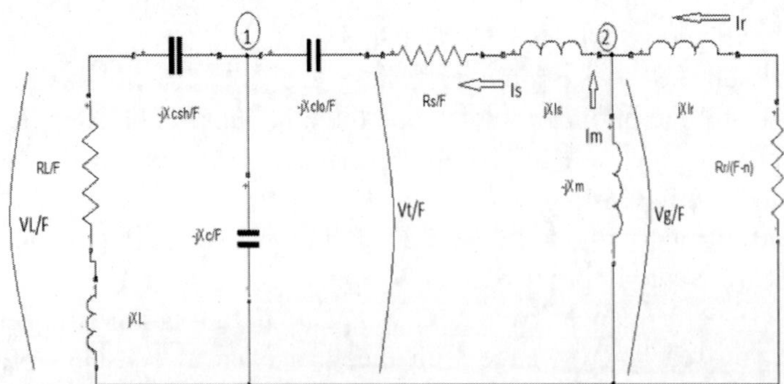

Figure 2. Equivalent circuit diagram of SEIG.

In Figure 2, Rs & Xs are the stator side resistance and reactance respectively, whereas & are the rotor side resistance and reactance respectively. The objective for achieving optimum root mean square voltage regulation, defined in Equation (1) by Fobj1, and corresponding fitness (F2) has defined in Equation (2).

Equation (1) represents the root mean square error of the difference between the voltage at the load and rated voltage from no load (n = 0) to load points.

$$Fobj1(Csh, Csse, n) = \sqrt{\frac{1}{NP}\sum_{N=0}^{NP}(\frac{Vln-Vr}{Vr})^2} \quad (1)$$

where Csh is Shunt capacitance in micro farads $Csse$ is short shunt series capacitance in micro farad n is speed in per unit.

The fitness correspondingly to objective function 1 has been represented by Equation (2)

$$F1 = \frac{1}{(1+Fobj1)} \quad (2)$$

The objective mentioned in Equation (1) are subjected to various limitations and bound, which are described as below:

- The real and imaginary parts of the loop impedance matrix are used to derive equality constraints. and the limiting values of various inequality constraints are characterized as:

$$X_m^{mn} \leq X_m \leq X_m^{mx},\ F^{mn} \leq F \leq F^{mx},\ V_l^{mn} \leq V_l \leq V_l^{mx}$$

- Also, Bound limits of capacitance at a given speed are important for self-excitation of short and long shunt SEIG. Here bound limits are considered as below:

$$12 \leq Csh \leq 31\mu F \qquad 180 \leq Csse \leq 450\mu F \qquad 0.85 \leq n \leq 1.15pu$$

For smooth SEIG operations X_m^{mn} and X_m^{mx} are taken as 0.1p.u. and X_m^{uns} respectively. For avoiding the insulation stress and overheating I_S^{mx} and V_S^{mx} are taken as 1.12 p.u. F^{mn} and F^{mx} are taken as n and 0.98n pu. From the point of view of quality V_l^{mn} and V_l^{mx} are taken as 0.95p.u. to 1.1p.u., respectively.

Artificial Intelligence Techniques

Artificial Intelligence (AI) plays a major role in optimizing the shunt and series capacitances required for voltage improvement in SEIG by choosing the best optima. GSA and GA are two AI algorithms that have been employed to optimize capacitance values in this chapter.

Gravitational Search Algorithm (GSA)

The Principle behind the GSA technique is Newton's law of gravity and law of motion which consist of exploration and exploitation. Exploration aids in discovering the optimal optima for good solution, whereas exploitation aids in determining the minimum and maximum capacitance values. Agents must act as though they are objects, and their performances are dependent on their mass. The SEIG parameters are optimized using GSA and the Newton-Raphson (NR) technique.

Procedure to Be Followed for SEIG Operation

1. In accordance to the search constraints, generate a random population. for i = 1, 2,...A.
2. Evaluate the fitness of objective function using equation (2).
3. Initialization of gravitational constant G0 used in Newton law of gravity which will update the best(t) and worst(t) of population

$$G(t) = Goe^{-\delta(\frac{t}{T})}$$

δ is a constant and T=Number of iterations.

4. Calculate masses, velocity and required acceleration for different location at different direction: Mi(t) is the gravitational and inertial masses

$Mi\ (t) = fiti\ (t) - worst(t) / best(t) - worst(t)$

where $fiti\ (t)$ is the fitness of agent ' i' at 't' iteration.

5. Update the velocity and positions of agents depending on random velocity as well as on acceleration.
6. Update new positions at the end of each iteration.
7. Repeat the steps until it reaches its optimum solution.

Genetic Algorithm (GA)

GA was the research on Charles Darwin's natural evolution hypothesis. The search and optimization of this algorithm were based on the notion of natural selection. Roulette-wheel selection, single-point crossover, and mutation operators are used in binary GA. Agents must act as though they are objects, and their performance is determined on their mass. The SEIG parameters are optimized using GSA and the Newton-Raphson (NR) technique. The constraints are required to compute and are solved using the NR method. The steps to optimize the capacitance and fitness values using GA has been described below.

Steps to Be Followed for Genetic Algorithm in SEIG

1. Initialize the GA parameters and set maximum iteration as itrmx, itrmnc as maximum iteration at which fitness is not changing and itrnc = 0 is not changing fitness iteration respectively.
2. Calculate Csh and $Csse$ for specified n and other equality, inequality and bound constraints.
3. Generation of the population of strings and compute Csh and $Csse$.
4. Using the fitness equation, calculate the fitness value for each string and $gbest = fitmax$.
5. Create the pool using Roulette-wheel selection while
6. Choose two strings from the pool, as well as the crossing location. Using a crossover procedure, create new offspring.
7. Carry out the mutation procedure on each bit of progeny created through the crossing.

8. Evaluate Csh and $Csse$ and compute the fitness of offspring. Set $itr = itr + 1$.
9. If $fitmax = gbest$, $itr\ nc = itr\ nc + 1$ otherwise gbest =fitmax, $itr\ nc == 0$.
10. If $itr < itrmx$ or $itr\ nc < itrmnc$ then go to Step 5, otherwise stop.

Result and Discussion

This study was conducted on a 3-phase, 5 kW IM that was used as a SEIG. Here, a 5 kW rated capacity were represented by 50 load points at rated voltage. The optimum performance using proposed optimization is implemented in the MATLAB environment. The simulation results are carried out for R load and 0.9 pf R-L load. Figure 3 represents the characteristics of short shunt SEIG feeding R load. Figure 3 shows the close argument between voltage and current levels.

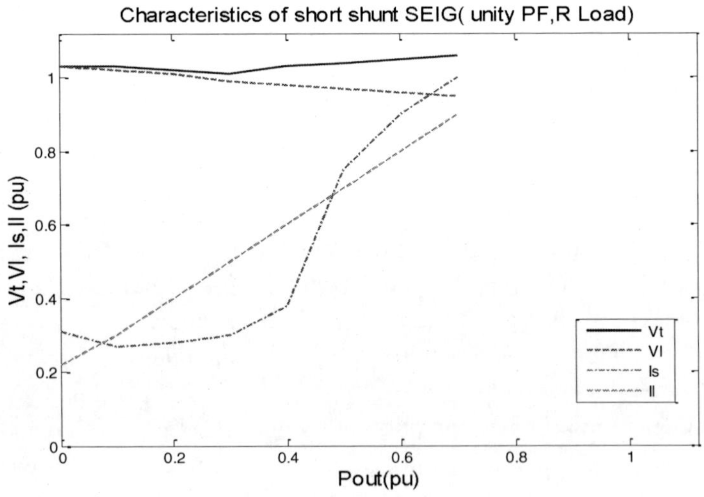

Figure 3. Characteristics of short shunt SEIG for R load.

Similarly, Table 1 represents the comparative analysis of GSA and GA for optimum voltage performance. Table 1 shows that an optimum power output of 0.779 per unit and fitness of 0.999 can be reached with a minimal shunt capacitance of 16.26 microfarad. In figure 4 the simulation results of voltage at different speeds for Csh = 17.2 microfarads and Csse = 410.2 microfarads. The curve is obtained for balanced resistive load using GSA and GA optimization techniques.

Table 1. Voltage regulation of short shunt SEIG under R load

n(speed)in pu	Fl	Csh in microfarads	Csse in microfarads	Pout in pu	Configuration
1.03	0.9999	16.26	390.6	0.779	GSA
1.08	0.985	11.28	427.7	0.772	
1.041	0.990	17.41	394.2	0.773	GA
1.09	0.990	14.31	421.0	0.770	

Table 2 represents the comparative analysis of GSA and GA for optimum voltage performance for R-L load. It has been investigated from Table 2 that at 29.88 microfarads of shunt capacitances optimum power output of 0.75 per unit and fitness of 0.998 has been achieved. The simulated results depict the close argument between the GSA and GA algorithm in terms of voltage regulation.

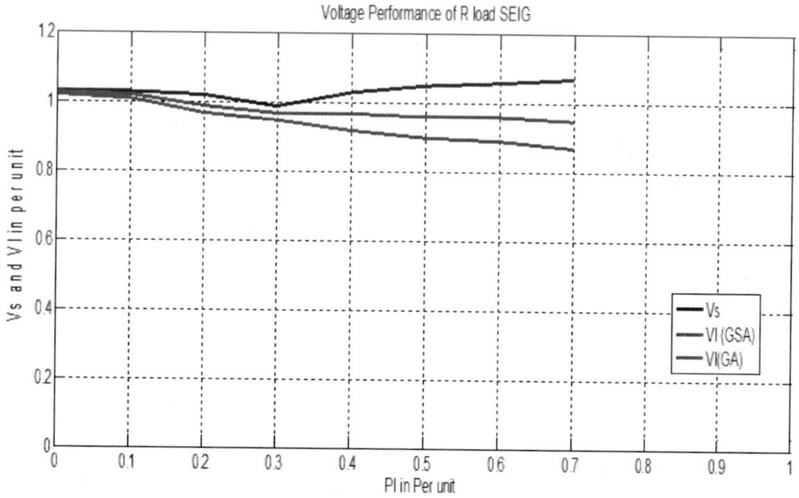

Figure 4. Voltage performance of SEIG for R load.

Table 2. Voltage performance of short shunt SEIG under R-L Load

n(speed)in pu	Fl	Csh in microfarads	Csse in microfarads	Pout in pu	Configuration
1.05	0.947	14.24	227.2	0.76	GSA
0.92	0.998	29.88	200.8	0.75	
1.071	0..975	13.86	199.7	0.744	GA
0.84	0.98	27.20	175.3	0.740	

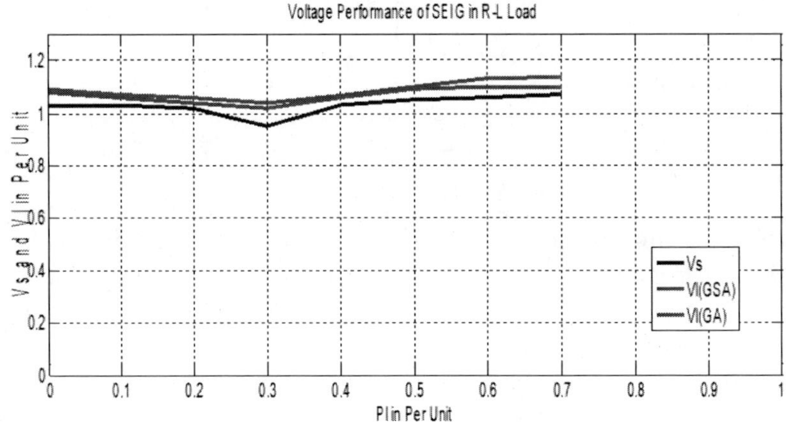

Figure 5. Voltage performance of SEIG feeding 0.9 pf R-L load.

Conclusion

In this chapter, the optimum performance of the SEIG machine has been formulated based on optimized shunt and series capacitance values. The optimization had performed for R and R-L load. The optimum Root Mean Square Error of voltage regulation has been considered as an optimization of objective function and solved using GSA and GA techniques. The closed agreement has been observed from both optimization techniques.

Appendix

Induction motor Parameters:

Three phase, 5kW, 415V, 10.2A(line),Rs = 0.072pu, Rr = 0.018pu,

$Xm^{uns} = 3.41pu$.

Prime-mover parameters:

Three phase, 10kVA, 1500 rpm, Shunt motor.

References

[1] Murthy S. S, Singh B and Sandeep V., (2016). Experience in developing a single phase two winding 5kW Self excited Induction Generator for Off Grid Renewable Energy Based Power Generation. *Journal of the Institution of Engineers (India) Series B*. 97(2), 127-137.

[2] Singh B, Murthy S. S, Gupta S. (2005), An electronic voltage and frequency controller for single phase self-excited induction generator for pico hydro applications, *IEEE PEDS*, 240-245.

[3] Haider A. M. A, Senan Md. F. M, Noman A, Radman T. (2012). Utilization of pico hydro generation in domestic and commercial loads, *Renewable and sustainable energy reviews*.16, 518-524.

[4] Faisal Khan Md, Rizwan Khan M and Iqbal A., (2017). Modeling, implementation and analysis of a high (six) phase self-excited induction generator. *Journal of Electrical Systems and Information Technology*, 156, 1–18. *Voltage Improvement of Short Shunt Self-Excited Induction Generator Using Gravitational Search Algorithm and Genetic Algorithm 11.*

[5] Aberbour A, Idjdarene K and Tounzi A., (2020). Performance analysis of self-excited induction generator mathematical dynamic model with magnetic saturation, cross saturation effect and iron losses. *Mathematical modelling of engineering problems, international information and engineering technology association*, 7(4), 527-538.

[6] Aly A, Aziz A, Hamdy R. A and Khalik A. (2019). Design and performance evaluation of a three phase SEIG feeding single phase loads. *Electric power components and systems*, 1-16.

[7] Wang Li, Huei, C. (2000). Long shunt and short shunt connections on dynamic performance of a SEIG feeding an induction motor load. *IEEE transactions on energy conversions,* 15(1), 1-7.

[8] Katre S. S, Bapat V. N, Induction generator for pico-hydro generation as a renewable energy source, *international conference on energy systems and applications (ICESA)*, 130-134.

[9] Rana K, Meena D. C. (2018), Self-excited induction generator for isolated pico-hydro station in remote areas, *IEEE International conference on power electronics, intelligent control and energy systems*, 821-826.

[10] Saha, S.K, Sandhu, K, S. (2018). Optimization techniques for the analysis of self-excited induction generator. *6th International conference on smart computing and communications ICSCC,* Elsevier, 405-411.

[11] Hannan M.A, Ali J, A, Azah Md, Hussain A. (2018). Optimization techniques to enhance the performance of induction motor drives: A review. *Renewable and sustainable energy reviews*, (2):1611-1626.

[12] Chauhan Y. K Jain, S. K, Singh B., (2011). Genetic Algorithm-Based optimum performance of compensated self-excited induction generator. *International journal of modelling and simulation*, 31(4), 263-270.

[13] Ibrahim H, Metwaly M (2011). Genetic Algorithm based performance analysis of self-excited Induction Generator. *Science Research Engineering*. 3(8), 859-864.

[14] Paliwal S, Sinha S. K, Chauhan Y. K. (2019). Gravitational search algorithm-based optimization technique for enhancing the performance of self-excited induction generator. *International journal of system assurance engineering and management*. 10,1082-1090.

Chapter 4

Micro/Pico Hydropower Generation System Using Self-Excited Induction Generators and Applications of AI for Its Performance Improvement

**Yudhishthir Pandey, Arif Iqbal,
Surya Prakash Singh and M. Aslam Husain**
Rajkiya Engineering College, Ambedkar Nagar (UP),
Uttar Pradesh, India

Abstract

Renewable energy resources (RES) are in high demand due to their advantages of a clean environment and non-perishable resources. However, RES faces drawbacks of intermittency and variations. Out of available renewable sources, micro/pico hydro has promising possibilities as RES. In this chapter, a detailed analysis of micro/pico hydro systems presented with aspects such as components used particularly, separately excited induction generator usage and application methods with improved voltage regulation, estimation of hydropower generation, and use of Artificial Intelligence (AI) techniques to improve the performance of the hydro system. This chapter focuses on performance improvement AI in planning, prediction of hydropower generation, operation and maintenance.

Keywords: hydropower, self-excited induction generator, machine learning, ANN, ANFIS, fuzzy logic

In: Applied Artificial Intelligence (AI) to Green Power Technology
Editors: Yogesh Kumar Chauhan, Ranjan Kumar Behera and Asheesh K. Singh
ISBN: 979-8-88697-131-6
© 2022 Nova Science Publishers, Inc.

Introduction

The increasing demand for electrical power with a cleaner environment has compelled innovators to think about distributed generations. Among the distributed generation or Renewable Energy Sources (RES), there are many options such as solar, wind, pico/micro hydropower etc. Low-cost generation using renewable power would lead to harness resources (Mathur, 2004). The significant advantage of these options is their easy installation and cheap electricity. However, there are many issues with these power options, such as variability and intermittent energy source with solar, wind and pico/micro hydropower (Kumar and Saini, 2021). A hydropower installation with a capacity of 5 kW to 100 kW is called micro-hydropower, and installation below 5 kW is categorized as pico hydropower. For a small (pico/micro) hydropower, the Self-Excited Induction Generator (SEIG) was found suitable with intermittent resources (H. C. Rai, A. K. Tandan, S. S. Murthy, B. Singh 1993, Arrillaga and Watson, 1978). SEIG performance is better than alternator for alternative energy sources such as wind power generation or micro/pico hydropower generation because of cost efficiency, brushless, without dc source need, overload and short circuit protection, and low maintenance. However, SEIG faces the disadvantage of poor voltage regulation when driven by fixed speed prime mover. The performance of SEIG is enhanced by using controllers (Rana and Meena, 2018).

A pico/micro hydro with natural resources and minimal civil work is a promising distributed generation solution for meeting the demand for electrical power in remote regions.

In this chapter, the authors presented a detailed discussion about pico/micro hydropower from installation to operation with applications of Artificial Intelligence (AI) to improve the performance of micro/ pico hydropower systems.

Commonly used AI-based techniques are Heuristics, Support Vector Machines (SVM), Artificial Neural Networks (ANN), and Markov Decision Process (MDP). These techniques used to make decisions based on extensive input data are efficient and accurate. AI finds application in discharge prediction, feasibility study, prediction of energy generation, and maintenance planning. Section IV will present a detailed discussion on each technique and application.

Further portions of the chapter organized as section II describe the various essential parts of micro/pico hydropower plant systems with the recent reorganization and innovations. Section III tells a brief overview of SEIG, describing its performance. Section IV describes the estimation of hydropower generation with the mathematical formulation. Section V describes the various techniques of AI for the performance improvement of micro/pico hydropower systems.

Description of Micro/Pico Hydropower Generation System

A typical layout of a hydroelectric power plant is shown in Figure 1. This hydroelectric plant consists of components such as a water reservoir, penstock, turbine, and power generation house.

Micro-hydro system operates as a 'run-of-river' system; that is, when running water flowing through the river passes through the turbines and generates electricity, after that, water is diverted back to the river/ stream with relatively low impact on the surrounding ecology. Generation of power ranges from 5kW to 100kW and up to 5kW are micro and pico hydropower generating system, respectively.

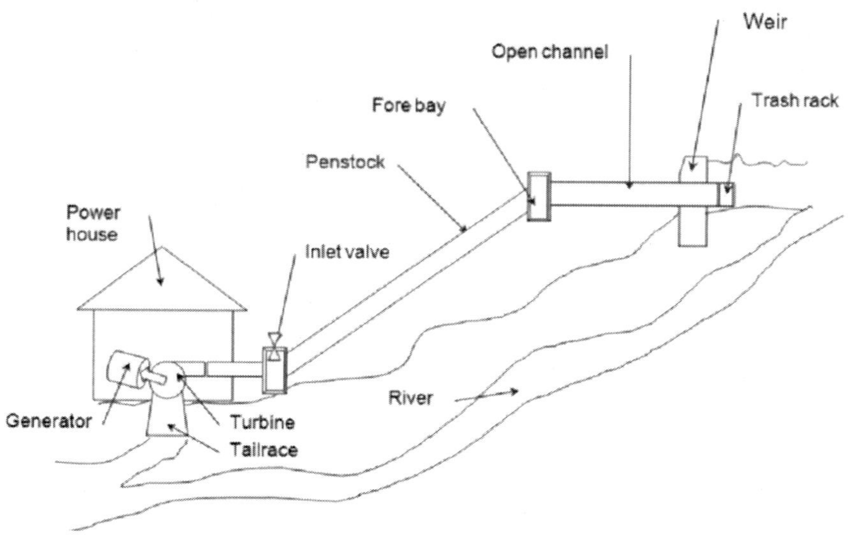

Figure 1. Layout of micro/pico hydropower generation system.

Control Gate or Flood Gate: Control or flood gates control the amount of water flowing from the river to the turbine through the penstock. The maximum volume of water passes to the penstock when the gate is wholly open, and water does not face any hindrance.

Penstock: A penstock made of steel delivers water from a river to a turbine through a channel, pipeline, or pressurized pipeline. Under gravity, water from height with stored potential energy flows down, passing through the penstock, converting potential energy into kinetic energy.

Run-of-the-river micro/pico hydropower systems consist of these essential components:

Water Turbine: Water reaches to water turbine through penstock with kinetic energy with the driving force. Now, the turbine is coupled with the rotating part of the generator. Now energy stored in water in the form of water flow drives the turbine. This result in electric power generation. Two types of the turbine are used based on water heads. First, Impulse turbine for large heads (sentence fragmentation). Second, a Reaction turbine for low and medium water heads.

Generator: A self-excited induction generator placed in the power evacuation house (sentence fragmentation). The turbine shaft is coupled to the rotating part of the generator. Blades of turbine receives water coming through penstock result it rotates, hence rotation of the rotor of SEIG and produces electricity. This generated electricity with the help of suitable converters such as electronic load controllers, improves voltage regulation (Bhim Singh and Tandon, 2002; Bhim Singh, Madhusudan Singh, 2010). This converter feeds a dump load and causes the match between the sum of the consumer load to the total power supplied by the generator with frequency and voltage within the prescribed limit (Mishra and Tiwari, 2017; B. Singh, Murthy, and Gupta, 2003; Bhim Singh, Murthy, and Gupta, 2006; Ramirez and Emmanuel Torres, 2007).

Self-Excited Induction Generator: An Overview

For renewable energy generation, a self-excited induction generator has emerged as a good option compared to a conventional alternator. For remote locations, an isolated SEIG mechanically coupled to renewable energy sources such as wind or micro/pico hydro turbines can supply agricultural and domestic loads where grid availability is possible.

In recent years with the extensive penetration of renewable energy, induction generator has dominated synchronous generator because of many advantages such as brushless, cost efficiency, robustness, protection against overload and short circuit and the capability of generating electrical power at a wide range of speed. An induction generator needs an external supply to produce a rotating magnetic field. The grid is known as a grid-connected induction generator if it supplies this external electrical power. If a separate capacitor bank provides this excitation, this arrangement is called a self-excited induction generator (Chan and Lai, 2002; Harrington and Bassiouny, 1998; Chan, 1993).

The major drawback of SEIG is poor voltage regulation as load increases. It does not maintain constant voltage as with the loading of the generator. Many methods have been proposed using FACTS devices such as Static VAR Compensator (SVC), STATCOMs (Chauhan, Jain, and Singh, 2010; H. C. Rai, A. K. Tandan, S. S.Murthy, B. Singh, 1993). However, these methods have difficulty of low reliability, complexity, and increased harmonics.

To improve the regulation of SEIGs, ELCs (a dummy load) to maintain constant load has been proposed (Rana and Meena, 2018). However, ELCs were power electronic controllers that increased power consumption by injecting harmonics. Future research is possible in this area to enhance the voltage regulation of the SEIGs.

Problem Formulation

In hydropower generation, potential energy stored in water at a height is used to generate electricity (Holland, 1983). In micro/pico hydropower generation, water with stored energy in the form of potential energy at a height 'h' meters reaches turbine blades. It transfers energy to the shaft based on flow rate 'Q' and efficiency. Taking flow rate 'Q' in litres per second passing through the penstock and reaching to turbine from a head 'h' in meters, hydropower is expressed as,

$$\text{Hydropower generated in kW} = k.h.Q \tag{1}$$

Values of k for Q and h in different units are shown in the attached table.

Here, the conversion efficiencies of different parts are taken as: turbine and generator combined with 60%. Usually, small turbines and generators are 75% and 80% respectively.

Hydroelectric power is generated from water stored at a height (with potential energy stored) which drives the turbine to rotate the coupled generator rotor. Energy extracted from the water depends on the volume of flow of water and speed of flow. The water reaches the turbine through the penstock, as shown in Figure 2. The height head difference between inlets to penstock to the outlet of penstock plays a vital role in the decision of high head because the kinetic energy of water hitting the turbine is proportional to the heads.

Table 1. Units and values of different quantities in hydropower generation

Units		
Q	h	k
Cubic foot/second	foot	0.05
Liters/second	Meters	0.006

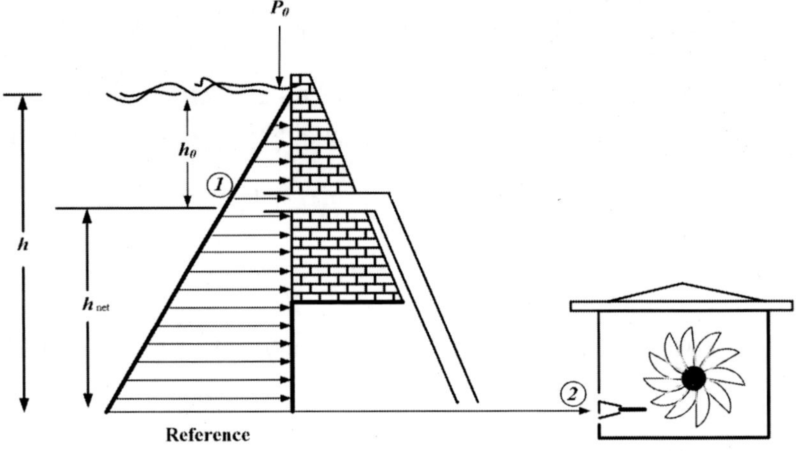

Figure 2. Water head and flow rate to turbine.

We can observe that the net pressure on the water at the top of the tank will be atmospheric pressure (Po), as shown in Figure 2. As we go deep into the tank, the pressure keeps increasing with linear variation with height (h). Therefore,

$$P=Po+\sqrt{}*h \quad (2)$$

where, P is the pressure at height h, √ is the specific weight of water, and h is height. Conservation of energy is the primary law behind the determination of hydraulic power.

Estimation of Hydro Capacity

Any system's power rating is the capacity to produce energy per unit of time. Units of power is Watt (W) or kilowatt (kW) and unit for energy is Watt-second (W-s) or kilowatt-hour (kWh). The power produced by micro/pico hydro systems mainly depends on two parameters. First, the head of the water is the height difference between the inlet and outlet of the penstock. Second, the water flow rate is available at the location of the hydropower plant. As we know, micro-hydropower plants have a power capacity of 5kW to 100kW, and pico hydropower plants might produce from a few watts to up to 5kW. Even 5kW capacity is well enough to supply the demand of two average Indian homes at remote locations. Hydropower capacity may be evaluated by the amount of water reaching the turbine (mechanical power input). Therefore, if m is the mass of water flow rate to turbine and h_{net} is the water head as shown in Figure 2. The water head defines the energy stored with water as potential energy. This potential energy is converted into kinetic energy (Energy conservation law) and forces the turbine to rotate. Consequently, the SEIG rotor connected with the turbine begins to rotate at speed higher than the synchronous speed of the rotating magnetic field. Based on the two parameters, the power output may be mathematically expressed as,

$$P = m*g*h_{net} \times \eta \tag{3}$$

In equation (3), g is the earth's gravity with $9.81 m/s^{2}$, and h_{net} is multiplied with a term (1-x), where x is the head loss. Specifically, If x is 10%, the $h_{net}=h_{gross}*0.9$. The η is the overall efficiency of the hydro system. Therefore, η is expressed as the product of the turbine, prime mover and generator efficiencies.

For micro/pico hydro systems, the turbine efficiency varies from 80 to 90% (taking 85%), prime mover efficiency is in the range of 90-95% (assuming 90%) and considering the generator's efficiency as 85%. The overall efficiency of the hydro system (η) shall be 0.85*0.9*0.85= 0.65025, i.e., 65.03%. Therefore, if we are considering a low gross head of 2meters for a micro/pico hydro system and a maximum flow rate of 3 m^3/sec, the power output (maximum) is estimated as follows:

Step-1: Net head will be the multiplication of gross head and 0.9 (considering 10% as a head loss)

$h_{net} = h_{gross} \times 0.9 = 2.0 \times 0.9 = 1.80$ m

Step-2: Maximum flow rate given in m^3/sec shall be multiplied by 1000 to get the flow rate in litres per second, so:

3 m^3/s = 3,000 litres per second

Step-3: Estimation of maximum hydropower:

Power (W) $= m \times g \times h_{net} \times \eta$

$= 3{,}000 \times 9.81 \times 1.80 \times 0.6503$

$= 34{,}448.99$ W = 34.449 kW

Now, if we estimate the maximum hydropower for a site with a high gross head and low maximum flow rate, then the estimated power will be close to the earlier estimated case for the same system.

The above illustration shows that two variables, head and water flow rate, play a vital role in estimating the generated output power of a micro/pico power plant. The output power is proportional to the head and water flow rate product.

The head level and flow of hydro resources are site-dependent. Based on these two parameters, a suitable turbine shall be chosen. For example, a low head site with a high flow will go for the Kaplan turbine. Kaplan turbine has a large dimensions, so a large powerhouse to accommodate and handle the large volume of water would be required. A high head site with a smaller flow rate will choose a small Pelton or Turgo turbine with the relatively smaller powerhouse size. This may be concluded that for micro/pico hydropower systems, site selection plays a vital role before installation.

Application of AI for Performance Improvement

The increasing demand for renewable energy sources (RES) has posed a challenge to grid stability due to the intermittent uncertainty involved with renewable sources. Power system requires more flexibility to operate with stability (Akrami, Doostizadeh, and Aminifar, 2019). With the advent of digital technologies in the energy sector, such as Blockchain technology (Andoni et al., 2019) Internet of things (IoT) (Bedi et al., 2018) with the feature of real-time control predictions based on data available has become more accessible. These predictions are accurate and precise. These technologies also help a safe, reliable, more effective, resilient and sustainable power system operation. Blockchain is distributed record managing set-up that attracted the attention of various stakeholders in the energy sector. Blockchain technology can be applied in the energy sector related to the operation and business processes such as automated billing (CIRED, 2017 2020), sales and marketing, trading and markets, automation, innovative grid applications and data transfer, grid management, security, sharing of resources, competition and transparency ("Blockchain in the Energy Transition. A Survey among Decision-Makers in the German Energy Industry | ESMT Knowledge" n.d.; "Blockchain in Energy and Utilities — Indigo Advisory Group|Strategy, Technology and Innovation" n.d.).

Internet of Things (IoT) is an essential part of an intelligent power system network. IoTs employed in power system networks include Supervisory Control and Data Acquisition (SCADA) and Advanced Metering Infrastructure (AMI). Deployment of IoT advantages are as follows:

- Improved reliability, adaptability, resiliency, and energy efficiency
- Enhanced sensing capabilities
- Enable on-demand information access and end-to-end service provisioning

Artificial intelligence (AI) is used for multi-purpose work in various areas such as economics, signal processing, control theory, optimization etc. AI involves a decision based on supervised and unsupervised learning. In supervised learning, decisions are based on past input and trained models. Unsupervised learning decisions are based on input, action, and reward. The application of supervised or unsupervised learning is decided by the nature of the applications. AI finds application in micro/pico hydropower for applications such as site selection (based on head and flow of water), feasibility observation, planning, power generation prediction, discharge prediction and overall maintenance planning. These applications are for an isolated pico/micro hydro site (Kumar and Saini 2021).

AI techniques can be classified as follows:

Machine Learning

Machine learning is a tool to analyze large amounts of data to solve various kinds of real-time problems. Output is obtained based on the input data analysis, learning, and training of multiple system models; hence, decisions can be ascertained. The output will be more accurate and precise, as we will have extensive data set as input for learning.

Machines learning has two types of learning (i) supervised learning and (ii) unsupervised learning—these two methods of machine learning are employed based on the application of the problem.

Deep Learning

Deep learning is a process used for inputs such as sound, images or text files. It deals with a large set of big data processing. Deep learning uses an artificial neural network with multi-layers to test and train the data set. Deep learning is an advanced machine learning technique with new variations. Deep learning utilizes an unbounded number of layers of finite size, which allows real-time practical applications. It implements by preserving theoretical universality under minor conditions.

Artificial Neural Network (ANN)

An artificial neuron network mimics the human brain, which takes various inputs with nodes and processes the inputs with activation function to produce output. A general layout of the ANN model is represented in Figure 3. Wherein x_1 to x_n are inputs multiplied by their weights (W_1 to W_n), generating the output ($x_1W_1+x_2W_2+ x_3W_3\ldots +x_nW_n$). This output is applied to the activation function to produce the final output. If ANN models involve more than two layers (one input layer and one output layer), those are intermediate layers. Intermediate layers are known as hidden layers.

The weight changes the strength of the signal at input neurons. Based on the threshold value signals are sent to process. Typically, neurons are aggregated into layers. Each layer processes several transformations on input signals (Senthil Kumar et al., 2013). ANN networks recognize patterns and split the data into abstraction layers.

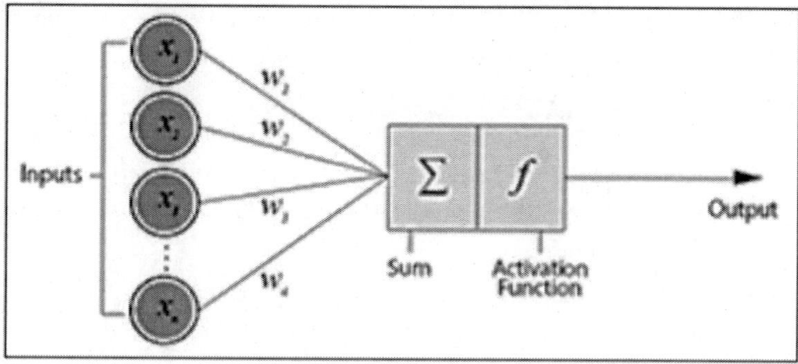

Figure 3. General layout of artificial neural network.

Fuzzy Logic

Fuzzy logic is a multi-valued logic that focuses on the "degree of fact" with values of variables, which may be any actual number from 0 to 1, unlike Boolean logic, which takes either 0 or 1 as input. This mathematical tool is a way to convert partial truth as input to map in output based on activation function or transfer function (Slany, 1994).

Fuzzy logic learns the data and produces output based on the abstraction layer.

Adaptive Neuro-Fuzzy Interface System (ANFIS)

ANFIS system combines ANN and fuzzy logic concepts to process the data. It is a kind of ANN which is the interface.

AI is used in the following applications: forecasting parameters, performance optimization, policy and feature selection, monitoring and control optimization, feasibility study, evaluation and capability assessment.

Conclusion

This chapter presents a study of the micro/pico hydropower system. In this study, a detailed discussion has been conducted on the need for a micro/pico hydropower system, components required for installation, power generation with a self-excited induction generator, estimation of hydropower generation, and application of AI to improve the performance of the system. This chapter concludes with the following points:

1. Remote locations power demand can be supplied with micro/pico hydropower system cost-effectively and reliably with the help of running rivers.
2. A SEIG can be used for micro/pico hydropower systems, as generators and regulations can be enhanced with modern controllers.
3. Hydropower generation is estimated with the head and flow of water.
4. Modern AI tools help in planning, site selection, power generation prediction, operation, and maintenance.

The presented study may be further expanded with the design of controllers for regulation enhancement of SEIG and better models for machine learning application in micro/pico hydropower systems.

References

Akrami, Alireza, Meysam Doostizadeh, and Farrokh Aminifar. 2019. "Power System Flexibility: An Overview of Emergence to Evolution." *Journal of Modern Power Systems and Clean Energy,* 7 (5): 987-1007. https://doi.org/10.1007/s40565-019-0527-4.

Andoni M, Robu V, Flynn D, Abram S, Geach D, Jenkins D, McCallum P and Peacock A. 2019. "Blockchain Technology in the Energy Sector: A Systematic Review of Challenges and Opportunities." *Renewable and Sustainable Energy Reviews,* 100 (October 2018): 143-74. https://doi.org/10.1016/j.rser.2018.10.014.

Arrillaga J and Watson DB. 1978. "Static Power Conversion From Self-Excited Induction Generators." *Proceedings of the Institution of Electrical Engineers,* 125 (8): 743-46. https://doi.org/10.1049/piee.1978.0177.

Bedi G., Kumar Venayagamoorthy G, Singh R, Brooks RR and Ching Wang K. 2018. "Review of Internet of Things (IoT) in Electric Power and Energy Systems." *IEEE Internet of Things Journal,* 5 (2): 847-70. https://doi.org/10.1109/JIOT.2018.2802704.

"*Blockchain in Energy and Utilities — Indigo Advisory Group*|Strategy, Technology and Innovation." n.d. Accessed January 14, 2022. https://www.indigoadvisorygroup.com/blockchain.

"*Blockchain in the Energy Transition. A Survey among Decision-Makers in the German Energy Industry* | ESMT Knowledge." n.d. Accessed January 14, 2022.

https://esmt.berlin/knowledge/blockchain-energy-transition-survey-among-decision-makers-german-energy-industry.

Chan TF. 1993. "Capacitance Requirements of Self-Excited Induction Generators." *IEEE Transactions on Energy Conversion,* 8 (2): 304-11. https://doi.org/10.1109/60.222721.

Chan TF and Loi Lei Lai. 2002. "Capacitance Requirements of a Three-Phase Induction Generator Self-Excited with a Single Capacitance and Supplying a Single-Phase Load." *IEEE Transactions on Energy Conversion,* 17 (1): 90-94. https://doi.org/10.1109/60.986443.

Chauhan YK, Jain SK and Singh B. 2010. "A Prospective on Voltage Regulation of Self-Excited Induction Generators for Industry Applications." *IEEE Transactions on Industry Applications,* 46 (2): 720-30. https://doi.org/10.1109/TIA.2009.2039984.

CIRED. 2017. 2020. "*Blockchain, Transactive Energy and P2P Trading Blockchain, Transactive Energy and P2P,*" no. June.

Harrington RJ and Bassiouny FMM. 1998. "New Approach to Determine the Critical Capacitance for Self-Excited Induction Generators." *IEEE Transactions on Energy Conversion,* 13 (3): 244-49. https://doi.org/10.1109/60.707603.

Holland R. 1983. *Micro Hydro Electric Power*. Intermediate Technology Publications.

Kumar K and Saini RP. 2021. "Application of Artificial Intelligence for the Optimization of Hydropower Energy Generation." *EAI Endorsed Transactions on Industrial Networks and Intelligent Systems,* 8 (28): 1-12. https://doi.org/10.4108/EAI.6-8-2021.170560.

Mathur HD. 2004. "Discussion of 'Bibliography on the Application of Induction Generators in Nonconventional Energy Systems.'" *IEEE Transactions on Energy Conversion,* 19 (3): 650. https://doi.org/10.1109/TEC.2004.832465.

Mishra Eshani and Sachin Tiwari. 2017. "Comparative Analysis of Fuzzy Logic and PI Controller Based Electronic Load Controller for Self-Excited Induction Generator." *Advances in Electrical Engineering,* 2017: 1-9. https://doi.org/10.1155/2017/5620830.

Rai HC, Tandan AK, Murthy SS, Singh B and Singh BP. 1993. "Voltage Regulation of Self-Excited Induction Generator Using Passive Elements." *Proceedings of the 6th IEEE International Conference on Electrical Machines and Drives, Conference*. https://doi.org/10.1109/T-AIEE.1935.5057024.

Ramirez JM and Torres ME. 2007. "An Electronic Load Controller for Self-Excited Induction Generators." *2007 IEEE Power Engineering Society General Meeting, PES,* 1-8. https://doi.org/10.1109/PES.2007.385540.

Rana Kailash and Duli Chand Meena. 2018. "Self Excited Induction Generator for Isolated Pico Hydro Station in Remote Areas." *2018 2nd IEEE International Conference on Power Electronics, Intelligent Control and Energy Systems, ICPEICES 2018*, 821-26. https://doi.org/10.1109/ICPEICES.2018.8897329.

Senthil Kumar AR, Manish Kumar Goyal, Ojha CSP, Singh RD and Swamee PK. 2013. "Application of Artificial Neural Network, Fuzzy logic and Decision Tree Algorithms for Modelling of Streamflow at Kasol in India." *Water Science and Technology,* 68 (12): 2521-26. https://doi.org/10.2166/wst.2013.491.

Singh B, Murthy SS and Gupta S. 2003. "Transient Analysis of Self-Excited Induction Generator with Electronic Load Controller (ELC) Supplying Static and Dynamic Loads." *Proceedings of the International Conference on Power Electronics and Drive Systems,* 1 (5): 771-76. https://doi.org/10.1109/PEDS.2003.1283000.

Singh B, Murthy SS and Gupta S. 2006. "Analysis and Design of Electronic Load Controller for Self-Excited Induction Generators." *IEEE Transactions on Energy Conversion,* 21 (1): 285-93. https://doi.org/10.1109/TEC.2005.847950.

Singh B and Tandon AK. 2002. "Dynamic Modelling and Analysis of Three-Phase Self Excited Induction Generator Using Matlab." *National Power Systems Conference,* 58-61.

Singh B, Singh M and Tandon AK. 2010. "Transient Performance of Series-Compensated Feeding Dynamic Loads." *IEEE Transactions on Industry Applications,* 46 (4): 1271-80.

Slany W. 1994. "*Fuzzy Logic in Artificial Intelligence*" 847. https://doi.org/10.1007/3-540-58409-9.

Chapter 5

An Investigation of Various Maximum Power Point Tracking Techniques Applied to Solar Photovoltaic Systems

Amit Kumar Sharma[1], PhD
and Rupendra Kumar Pachauri[2], PhD
[1]Department of Electrical Engineering, Government Polytechnic,
Kirthal, Bagpat, Uttar Pradesh, India
[2]Department of Electrical and Electronics Engineering, School of Engineering,
University of Petroleum and Energy Studies,
Dehradun, Uttarakhand, India

Abstract

Over the last few years, due to the shortcomings of traditional energy sources, there has been a rise in global interest in renewable energy (RE) sources such as tidal, MHD, geothermal, fuel cells, wind, bio energy & solar etc. for the power supply to isolated grids. RE sources of energy are clean, inexhaustible and environment friendly. If properly harnessed, these powerful energy sources can supply large grids/ isolated loads. Solar and wind energy are commonly used for this reason due to their abundant availability. However, these energies have certain limitations, such as high investment costs and poor efficiency. Solar-driven energy production systems, at the very first stage of conversion, depend on climatic conditions such as solar radiation, temperature, partial shade panels, etc. Second, solar photovoltaic (PV) must, under all conditions, provide sufficient power to meet the linked load requirement. To this end, several maximum power point tracking (MPPT) controllers through different tracking techniques are used to boost solar PV systems yield. However, at maximum power point (MPP), there is a compromise between accuracy and stability for each MPPT technique that influences

In: Applied Artificial Intelligence (AI) to Green Power Technology
Editors: Yogesh Kumar Chauhan, Ranjan Kumar Behera and Asheesh K. Singh
ISBN: 979-8-88697-131-6
© 2022 Nova Science Publishers, Inc.

the performance of PV systems. Also, the DC converters associated with the PV array output to buck/boost its output power also influence the system performance. It is therefore observed that there are several factors influencing the performance of PV system. In view of all these factors, the need for an hour is to develop an efficient system to increase overall efficiency for efficient use of solar energy under the various conditions mentioned above. In literature, several researchers are working to build successful models to solve these problems. Present work focuses on several MPPT techniques incorporated in solar PV systems for the extraction of maximum solar power at all times, developed by various researchers for the supply of grid or isolated DC loads.

Keywords: maximum power point tracking, photovoltaic system, artificial intelligence, solar energy

Introduction

Energy plays a vital job in the progression of any country as well as in its daily life (Sharma et al., 2018). Nuclear, Electrical, Mechanical and chemical are the basic forms of energy which are available to us. Due to the rapid growth in world population, energy demand in all countries is increasing day by day. Fossil fuels like natural gas, coal and petroleum are the basic sources of energy available and due to their increased consumption, they may get exhaust someday. This may lead to the situation of energy crises due to the shortage of energy. Thus, it is very essential for us to find the other sources of energy to meet prospect energy needs.

Since energy conservation law says that "Energy can neither be produced nor be shattered though it can be changed as of one type to a different", developed countries finds alternate means of extracting energy other than conventional sources of energy which are available in the form of coal, natural gas and petroleum. These sources are classified as non-conventional or renewable sources of energy. These sources of energy have huge advantages as compared to conventional sources of energy (Fernandez et al., 2019) but have some limitations too. Energy can be produced again and again from these sources as they are inexhaustible. Solar, wind, geothermal, tidal are some basic non-conventional sources of energy which are used to provide electrical energy demand in developed and developing countries (Benlahbib et al., 2020; Madaci et al., 2016). Since solar energy is readily available all over the world, more development is seen in this field in many countries (Bandaru et al., 2019;

Tina and Grasso, 2014). This chapter summarize the basic concept of solar energy; problems associated in extracting it and advance techniques which extract this energy efficiently.

Basics of Solar Energy

Sun is the vital supply of all energies on this planet. It releases huge amounts of energy by the process of thermonuclear fusion process. Sun radiates this energy in all directions (Zeitoury et al., 2018) of which only a small portion reaches to the earth surface. Nowadays research is going on in utilizing this extent of energy reaching the earth surface to meet the energy demand of human beings (Sharma and Puri, 2020). Researchers find solar cell which converts solar energy into electricity (Islama et al., 2018; Nkambule et al., 2020). With the advancement in technologies these cells are utilized in an efficient manner and form the basic element of solar PV system.

This cell is basically built by semiconducting material, usually silicon with thick p layer and thin n layer. These cells work on photovoltaic effect. When the photons from sunlight hit the plane of solar cell, it releases the pair of holes and electrons which results in flow of electric current through an external circuit. Thereby transforming solar energy into electrical.

A combination of solar cells in series is termed as solar module. Solar modules are connected to form solar panels and when this panel on large scale connected in series and parallel, they form solar array.

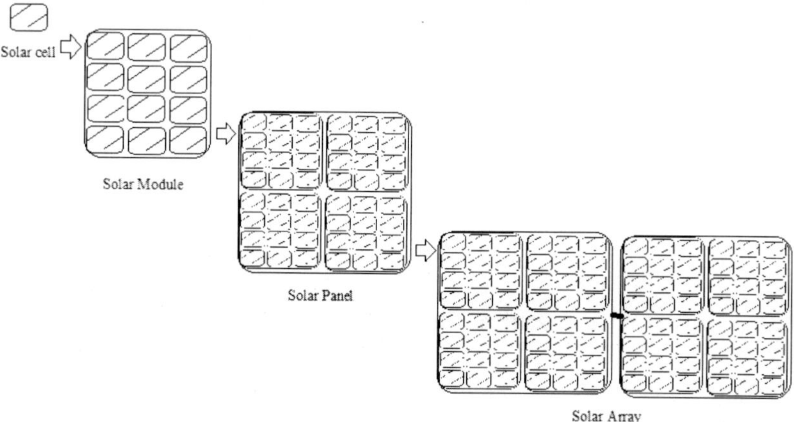

Figure 1. PV cell, module, panel and array.

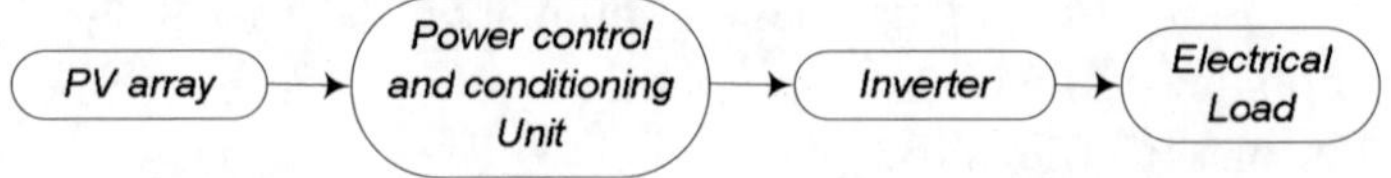

Figure 2. Elements of PV system.

While extracting solar energy into electrical energy these power plants may work either independently or in conjunction with the power grid (Solanki et al., 2017). While operating independently these plants are called autonomous power plant which is mainly used to supply the local network. If these power plants are used to supply external network, they are called grid coupled PV system. PV system schematic is shown in Figure 2.

Problem Encountered in Establishing Solar PV System

Though a lot of research and emerging technologies list the benefits of utilizing solar energy, one must be familiarized with the various problems encountered while installing a solar PV system. Some of them are outlined below:

- Efficiency of these systems depends on various factors; common among them are purity of the semi conducting material used, changing environmental conditions, Insulation resistance and formation of snow, ice and dust layers on panels.
- Installation of these systems requires large area due to the presence of control room, service room, monitoring room & storage room, which increases the initial installation cost of these systems.
- Problem in generating constant power due to change in atmospheric conditions & shading conditions of nearby obstacles
- Problem in giving maximum power from sunrise to sunset. To maximize the output at all time solar trackers are employed.

Solar Module Characteristics

As Output current of solar module depends on insolation level and its output voltage depends on temperature, main electrical features of PV cell/module can be depicted in its current-voltage *(I-V)* curve. This is basically a graphical illustration of solar cell/module operation at particular conditions of temperature and irradiance shown in Figure 3. Through this curve solar systems are built to drive at MPP at all time.

Open circuit voltage 'V_{oc}' in characteristics is obtained under no load condition in PV system i.e., load is disconnected from the module. 'I_{sc}' is the short circuit current that flows at the output of cell/module when its output terminals are short circuited.

In between these two points solar cell/modules can generate the maximum power at a one combination of current & voltage known to be MPP. It's usual practice to operate the cell/module on this point so that we can get maximum power at all time. But this point shifted in the characteristics as the values current and voltage is altered with the changing atmospheric conditions.

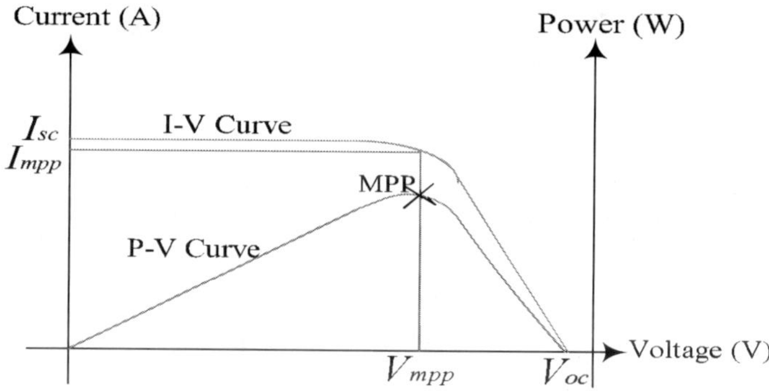

Figure 3. Solar module P-V & I-V characteristics.

A variety of techniques are discovered by many researchers for chasing this point during the working of solar cell/module and referred as MPPTs. These are the advanced methods through which one can extort utmost power from solar PV system at all time even during changing environmental conditions.

Maximum Power Point Tracking Techniques

MPPT techniques are integrated in PV system are based on adjusting the impedance seen by the solar panels to keep the system operation at a point where power obtained is maximum under any circumstances. These techniques are implemented with DC-DC converters so as impedance of source and load is matched by altering DC converters duty cycle. These techniques are broadly classified as online & offline techniques (Ali et al., 2020; Bastidas-Rodriguez et al., 2014; Bhatnagar and Nema, 2013; Subudhi and Pradhan, 2013; Gupta et al., 2016).

In online MPPT control strategies, generally instantaneous output electrical quantities of PV module are measured for generating control signal in order to accomplish MPP in changing environment circumstances. Generally, DC converters duty cycle is altered (Kumar et al., 2015) to achieve this through directly, P, PI or PID controller. In literature it has been seen that these techniques can track true maximum power point and their convergence speed depends on change in step size. Power oscillations with large step size and more time to track with small step size are the major limitations of these methods. Perturb and Observe (P&O), Hill Climbing, Increment Conductance (INC), Feedback control, Linear Current control, are some techniques which falls under online categories. Whereas offline control strategy makes use of measuring one or more electrical output quantity of PV panel by disconnecting it from load and forces operating point of panel towards maximum power point. These techniques have high convergence speed and are easily put into operation. There is a power loss while disconnecting the PV panel during measurement. Moreover, true MPP is not achieved by these methods, that's why it is not preferred where efficiency requirement is high. Fractional open circuit voltage (FOCV) and Fractional short circuit current (FSCC) techniques drop under the class of offline techniques. Some of the techniques used the benefit of both online and offline method and called hybrid methods. The following section explains various MPPT algorithms reported by various researchers to track MPP. Various optimization algorithms have been developed in conjunction with conventional techniques nowadays to achieve desired power point efficiently.

Perturb and Observe (P&O) Technique

Output power of PV module is regularly perturbed in this method to chase MPP at all time (Christopher and Ramesh, 2013). Solar PV system yield power is continuously monitored at regular intervals & compared with its preceding value. PV module voltage is perturbed to monitor the variation in PV module yield power. If a boost in voltage results in power enhancement, it shows that the PV module functional point is in the left region of MPP. Therefore, more perturbation is needed towards right side achieve MPP & Vice-versa. Change in yield voltage is achieved by altering DC converter duty cycle or step size used at the output of PV module. Figure 4 shows flow chart of this technique.

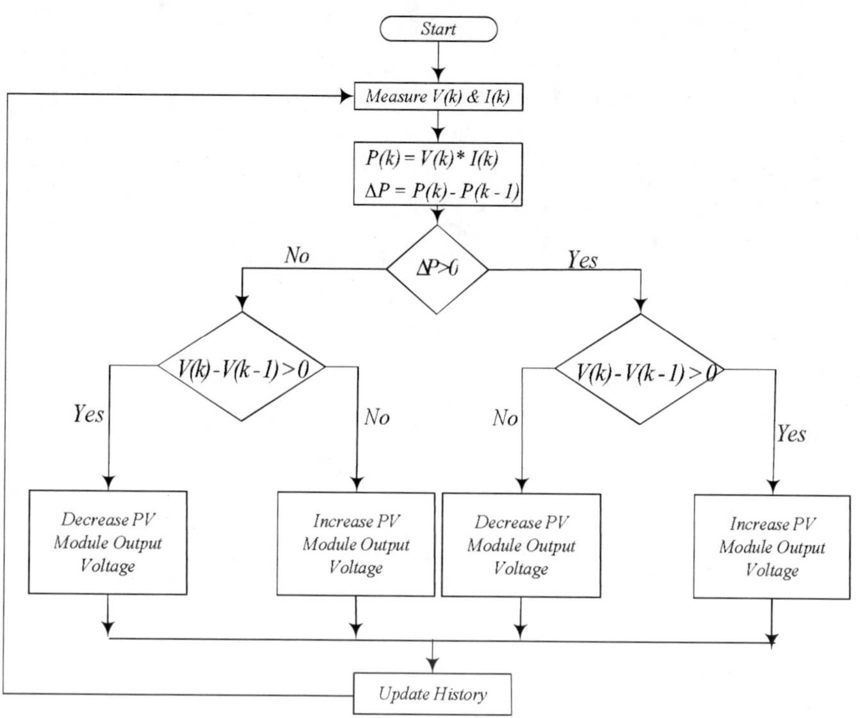

Figure 4. P&O technique (Christopher and Ramesh, 2013).

Process of perturbation and observation continues till MPP is reached. Main problem with this technique is fixing the step size value to track MPP. If the size of step is selected small, the system will show a slow response results in more time to track MPP. Alternatively, if the value of step size is chosen large, power loss occurs due to inaccurate tracking. So, choosing the correct value of step size is the utmost important parameter of this technique. (Femia et al., 2007; Femia et al., 2005; Alik et al., 2015) reports the improved P&O techniques having improved system response. Many optimization algorithms are included in this method; two among them are discussed next.

P&O Based Multiple Power Sample MPPT Technique

This approach is based on the concept of using multiple successive functioning points and inspection of output power variation between two successive points

(Abouadane et al., 2020). This will provide the direction of operating point with regard to MPP whether it is moving away or towards it. This algorithm checks the alteration in solar irradiance level by analysing and sampling the power between two consecutive points and eliminates its effect from the comparison. Change in power is now only due to perturbation given by algorithm. After comparison the algorithm perturbed the PV voltage to shift the operational point near MPP. Afterwards updating the value of reference voltage, this algorithm shifts one step back, the value of sampled power and voltage and ensures that the operational point is always near the MPP. Figure 5 shows flowchart of this method.

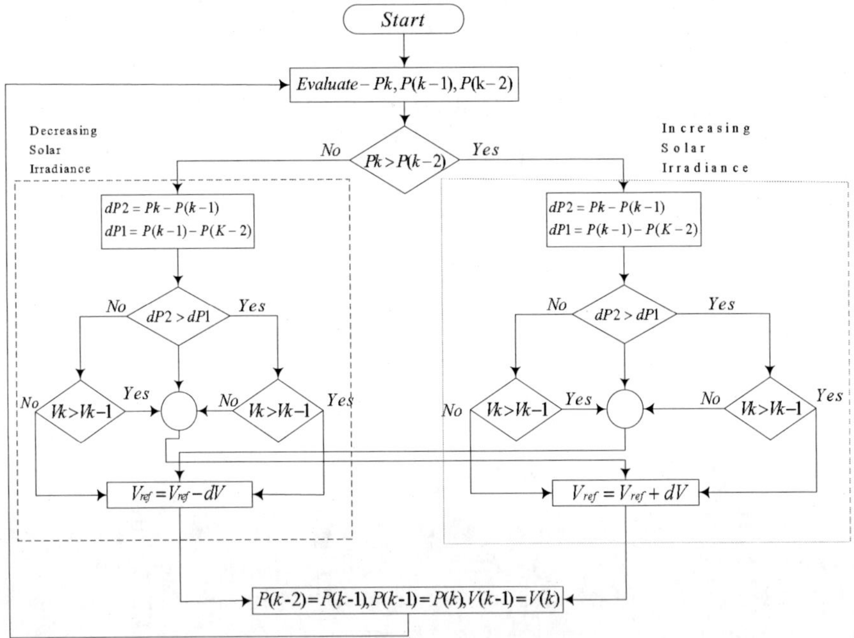

Figure 5. Flowchart - P&O based multiple power sample MPPT technique (Abouadane et al., 2020).

Adaptive Perturb and Observe Technique

This approach made use of simplified model-based state evaluation along through adaptive alpha P&O technique (Wang et al., 2020) which helps in increasing the tracking of MPP accurately. Algorithm has two stages to follow the utmost power accurately. In the first phase only two sampling points of voltage and current are required on panel operating characteristics for the estimation of temperature and irradiance which is based on simplified model-based state evaluation. In the later stage the operational point of panel is stimulated directly to operational voltage directive, which is obtained by fitting association between MPP voltages in different insolation levels. Afterwards alpha P&O strategy tracks the accurate MPP. In this stage alteration in power is used to determine variations in irradiance level. If change in irradiance occurs, algorithm executes two further perturbation steps & runs simplified model-based state estimation cycle to calculate new values of irradiance and operating voltage command. Figure 6 illustrates the flowchart of this algorithm.

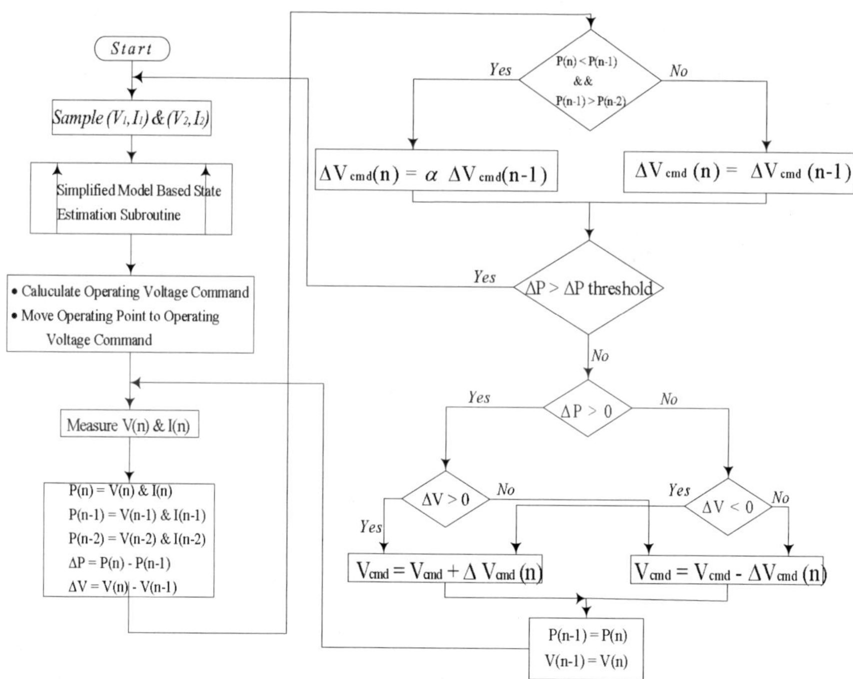

Figure 6. Adaptive perturb and observe technique (Wang et al., 2020).

Incremental Conductance Method

This strategy is selected for tracking MPP under fast changing environment conditions, which is major drawback of P&O method (Elgendy et al., 2013; Zakzouk et al., 2016). Module instantaneous conductance *(I/V)* is compared by its incremental conductance *(ΔI/ΔV)* for chasing exact MPP. Table 1 summarizes the various conditions of chasing exact MPP by this technique.

Table 1. Conditions in INC method to track MPP

	Position of operating point
$\left(\frac{\Delta I}{\Delta V}\right) > \left(-\frac{I}{V}\right)$	Left side of PV characteristics
$\left(\frac{\Delta I}{\Delta V}\right) < \left(-\frac{I}{V}\right)$	Right side of PV characteristics
$\left(\frac{\Delta I}{\Delta V}\right) = \left(-\frac{I}{V}\right)$	Is at MPP of PV characteristics

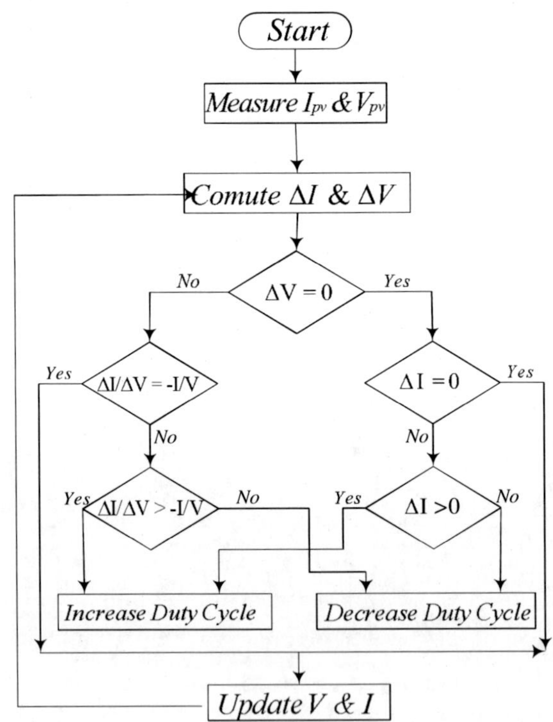

Figure 7. Incremental conductance algorithm (Mirza et al., 2020).

Basically, INC technique alters the duty cycle of DC converter incorporated in PV system to change the module output voltage till the condition *(ΔI/ΔV)* = *(-I/V)* is reached and hence found the MPP. INC algorithm can track exact MPP without oscillations around MPP as observed P&O technique. Moreover, this method has high accuracy while tracking in varying environment conditions. Main disadvantage of this technique is that noise may it's altered the performance (Freeman 2010), due to which it may continue to search for the condition $(\Delta I/\Delta V) = (-I/V)$ as shown in Figure 7 flowchart. Algorithm of the technique is complex than P&O method. Modern optimization techniques use this method to track the MPP more efficiently.

Regulated Incremental and Conductance MPPT Technique

As discussed above incremental and conductance method alters converter duty cycle for tracking MPP, but this conventional algorithm doesn't show high performance in both transient and steady state situations. This inadequacy is overcome by regulated incremental and conductance algorithm proposed in (Wellawatta and Choi, 2019). This method uses digital compensator to evaluate INC function which improves its power tracking ability. Author defines the INC function as

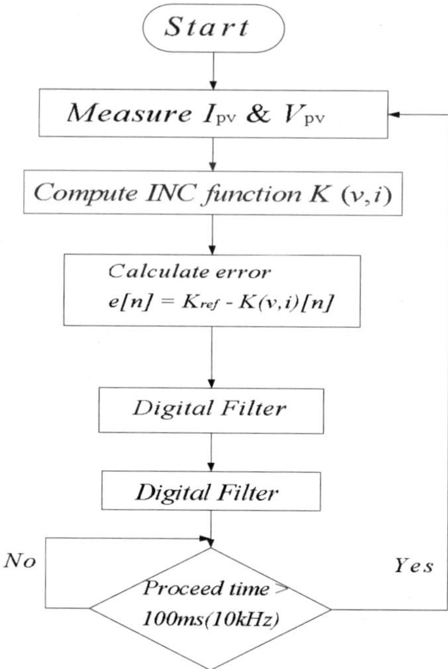

Figure 8. Regulated incremental conductance MPTT algorithm (Wellawatta and Choi, 2.2.).

$$K(v,i) = \frac{I}{V} + \frac{\Delta I}{\Delta V} \quad (1)$$

Smaller the value of this function, operational point is closer to MPP. Its value is zero at MPP. Proposed algorithm obtains an error signal, after comparing INC function at the existing & reference operating points. Furthermore, operating point is navigated by control loop to reduce error. The operational point gradually reaches the MPP if reference is selected as zero. Hence DC converter duty cycle is modified with digital filter in accordance with error direction as shown in Figure 8 flowchart. This will enhance both speed and stability of system while tracking maximum power point.

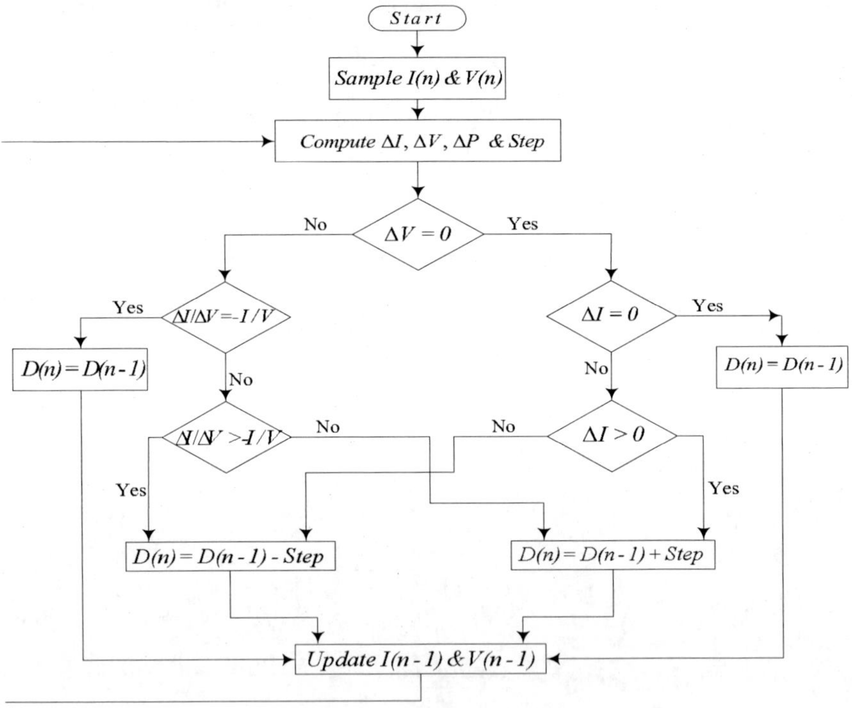

Figure 9. Variable step incremental conductance MPTT algorithm (Zakzouk et al., 2013).

Variable Step Incremental Conductance Technique

A new approach of INC technique adopting variable step is proposed in (Zakzouk et al., 2013). Conventional incremental conductance algorithm calculates the fixed step size keeping accuracy and tracking speed in mind.

Thus, there is a compromise between steady state & transient part of the structure. With variable step size approach, step size of the next iteration is automatically examined which enhances the output yield and tuned with the characteristics of PV module. If the operational point is located distant from MPP, algorithm enhances the step size which grades in high tracking speed. While it decreases the step size if the operating point is located near MPP, it results in decreasing the oscillations around MPP. Hence the effectiveness of the overall system is enhanced. Flowchart of this MPPT approach is shown in Figure 9. Variable step is defined by the author as

$$T(n) = T(n-1) \pm k\left(\frac{\Delta P}{\Delta T}\right) \quad (2)$$

where, k = scaling factor coefficient which mainly determines the system performance. Tuning of 'k' is determined in (Bastidas-Rodriguez et al., 2014). ΔT = Step change in duty cycle.

Fractional Open Circuit Voltage Method (FOCV)

FOCV is simplest offline MPTT technique demonstrated by many research works (Ahmad, 2010; Baimel et al., 2016; Frezzetti et al., 2014; Huang, 2014; Lopez-Lapena and Penella, 2012). Linear relationship between V_{oc} (Open circuit voltage) & V_{mpp} (Voltage at MPP) of PV module is the base of evolution of this technique.

$$V_{mpp} = k \times V_{oc} \quad (3)$$

where K is proportionality or voltage factor [35] depends on PV module *I-V* curve. It's typical range is 0.7-0.9 specified on PV module datasheet. V_{mpp} can be calculated by monitoring module V_{oc} periodically disconnecting the module from the load and multiplying it with '*k*' i.e., with voltage factor. With this a precise evaluation of V_{mpp} can be made because the value of '*k*' doesn't depend on changing environment conditions. After the estimation of V_{mpp}, closed loop control of the system makes the panel operate at this voltage and make it to deliver maximum power at all time. Main benefit of this scheme is its easy implementation and low cost. Due to regular measurement of V_{oc}, power converter is momentarily shut down results in loss of power. Various works have been carried out in this field to minimize this power loss in the last few decades. Pilot cell scheme is used to resolve this problem (Baimel et al., 2016).

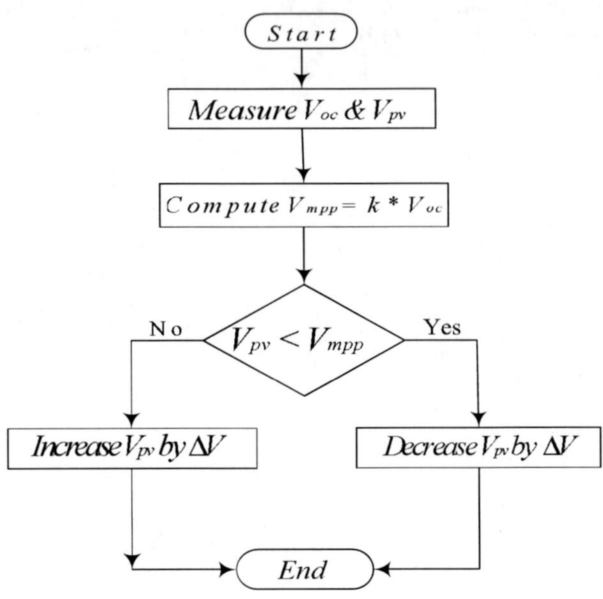

Figure 10. FOCV method (Baimel et al., 2019).

Pilot cell is the additional cell like PV module cells used only to measure open circuit voltage eliminating the necessity of load disconnection. This cell doesn't produce any power and is isolated from the photovoltaic module. Extension of this method is proposed in (Baimel et al., 2019).

Semi-Pilot Cell FOCV MPPT Technique

This research introduces a semi-pilot cell in place of pilot cell which PV module incorporate as its part and hence enhances total power generation. Semi pilot cell supplies the load as other module cells during supply period. This cell can be disconnected from the module for the measurement of open circuit voltage during measurement period. A switch is used to bypass the semi pilot cell in measurement period as shown in Figure.11. This scheme provides less power loss results in efficient utilization of pilot cell contrast to FOCV method.

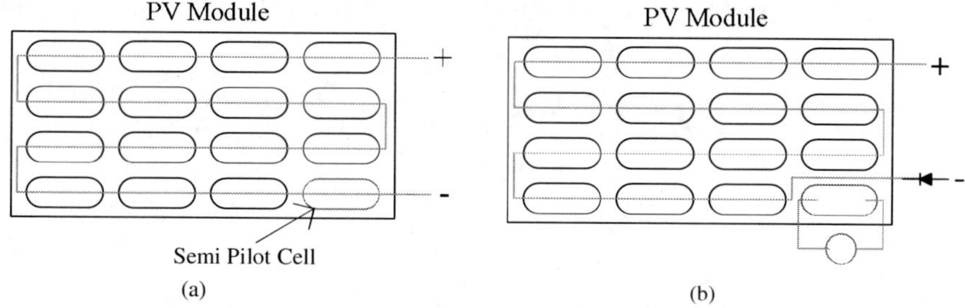

Figure 11. (a) All PV module cells supply power to load (b) disconnection of semi-pilot cell to measure V_{oc} (Baimel et al., 2019).

Fractional Short Circuit Current (FSCC) Method

This procedure utilizes the fact that ratio of current at MPP (I_{mpp}) to module short circuit current (I_{sc}) is fixed because I_{mpp} is linearly depends on I_{sc} (Masoum et al., 2002; Bekker and Beukes, 2004; Hart et al., 1984; Noguchi et al., 2002) as,

$$I_{mpp} = k \times I_{sc} \tag{4}$$

where '*k*' is constant generally ranges between 0.78- 0.92. I_{sc} is measured at regular intervals in continuously varying environments by shorting the module terminals. This value is then multiplied by '*k*' to obtain reference signal used by converter. Figure 12 shows the flowchart of FSCC method.

Measuring short circuit current requires an extra switch with power converter that shortens the PV module output terminals at regular intervals. Therefore, the cost and components of the system increase. Measuring short circuit current also reduces power consumption, thus true MPP is never reached.

It is seen that changing module characteristics creates problems with fractional measurement methods (FOCV & FSCC). When the module measures the fractional parameter, the controller must account for its real time parameter variations. Since operating conditions of module supplying the load is different from those permanently opened or shorted module, tracking of maximum power point may be inaccurate. Therefore, these techniques are hardly implemented as autonomous algorithms. They usually used with other advanced methods for reducing the time to reached at MPP.

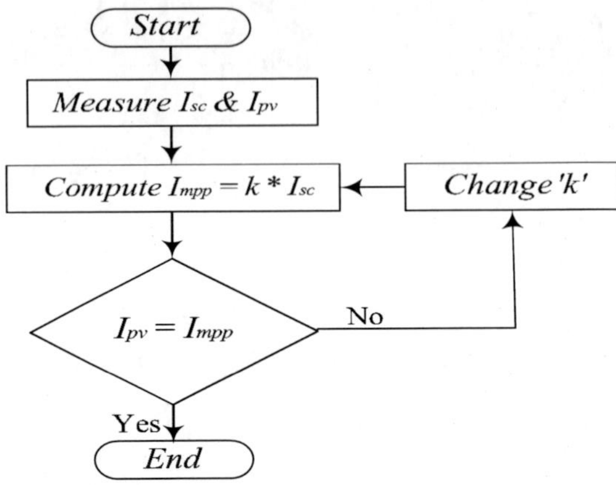

Figure 12. Fractional short circuit current method.

Soft Computing Techniques

Now-a-days implementation of soft computing techniques in designing MPPT controllers attracts many researchers (Salam et al., 2013; Mellit and Kalogirou, 2014; Lian et al., 2014; Ishaque et al., 2012; Gupta et al., 2014; Pachauri and Chauhan 2014). These techniques are more efficient in tracking MPP under any environment condition. If these techniques are combined with conventional MPPT algorithms (Alajmi et al., 2011; Firdaus et al., 2020; Chamanpira et al., 2019), shows superior performance of the system. Though soft computing techniques highly improve tracking time and precision, they suffer from the disadvantage complexity. Many microcontrollers are designed based on these algorithms. Various soft computing techniques are shown in tree below.

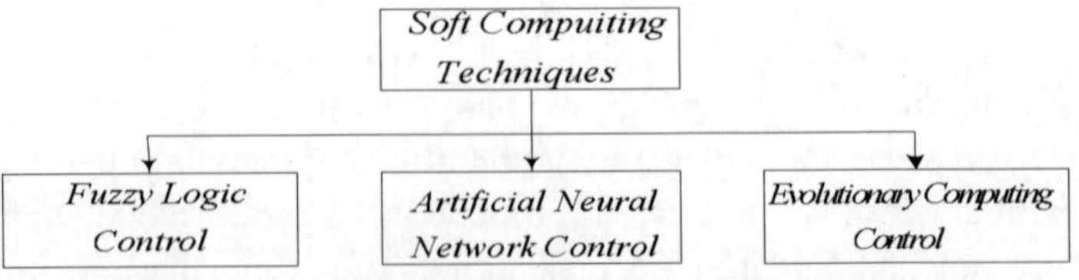

Figure 13. Categories of soft computing techniques.

Fuzzy Logic Control (FLC)

Controllers based on FLC exploit linguistic variables based on designer knowledge. These controllers can work with imprecise inputs and are able to deal with nonlinearity. An accurate mathematical model is not required by them. Figure 14 shows the structure of FLC (Hannan, 2009). Calculations are performed in rule interface block. MPPT controllers based on FLC usually have error '*e*' or alter in error 'Δe' as its inputs (Noureddine et al., 2015; Alajmi et al., 2011).

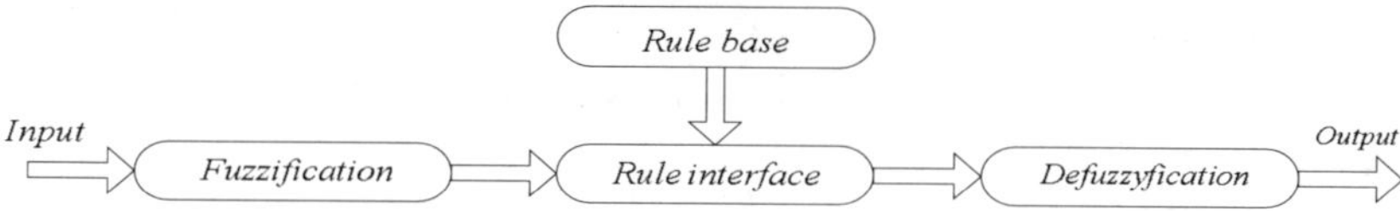

Figure 14. FLC block diagram.

Both the inputs can be calculated as,

$$e(n) = \left(\frac{P(n) - P(n-1)}{V(n) - V(n-1}\right) \tag{5}$$

$$\Delta e = e(n) - e(n-1) \tag{6}$$

where P & V represents power & voltage of PV curve. After evaluating '*e*' and 'Δe' membership function transformed them to linguistic variables (Bouchafaa and Boucherit 2010). Sometimes less symmetric membership functions are designed to provide more value to specific fuzzy level (Simoes et al., 1998; Masoum and Sarvi, 2005; Chian-Song, 2010). In defuzzification block, output linguistic signals are transformed in numerical values. Many methods are used to perform this task (Cheng et al., 2015); centre of gravity is one among them. One main drawback of fuzzy logic control based MPPT controller is that they cannot work efficiently in partial shading condition once their rule table is fixed.

Artificial Neural Network (ANN) Control

Another soft computing method is ANN shows significant improvement in tracking MPP (Hiyama et al., 1995; Ro and Rahman, 1998; Hussein et al., 2002; Zhang et al., 2002). ANN is an artificial neuron (node) interconnection that copies the natural brain. It consists primarily of input, secret and output layers. Each layer nodes may vary & define by user. PV module parameters, atmospheric irradiance or temperature can be the input variables. Depending on the control variables required by converter, the output may be duty cycle, current or voltage. ANN's ability to monitor the MPP depends on the hidden layer of optimization technique and how carefully and thoroughly the networks have been trained. Neurons are weighted correctly during the training to go with the relationship of input-output patterns. Since PV module characteristics vary with time, ANN must be trained at regular intervals to assure true MPPT.

Evolutionary Computing Control

Since traditional MPPT techniques suffer from the drawback of tracking MPP under partial shading conditions, researchers focus on soft computing techniques (Bingöl and Özkaya, 2019) due to their consistency and flexibility. An evolutionary computing control based on soft computing techniques is presented in Figure 15. Genetic algorithm (Bingöl and Özkaya, 2019) is based on the philosophy of natural progress. Aim of this algorithm is to make foremost species from its antecedent. Algorithm specifies that the candidate solution is a chromosome consisting of a fixed gene pool. Selection, crossover and mutation are its fundamental operators. A chromosome is arbitrarily chosen from the live population for addition in its fitness value in the subsequent generation during the selection process. While two chromosomes are fused in the crossover to form a fresh chromosome and genetic diversity is provided by mutation process.

Differential algorithm is another soft computing evolutionary control. This algorithm is centred on mutation, such that the operating point in the search region is closer to the optimum value. It also has three basic courses of action as selection, crossover and mutation (Bingöl and Özkaya, 2019). It starts with the initialization process and the initial parameters are randomly initialized by the population. Fitness function is calculated for all individual in the population, and the operational point is transferred to the most excellent solution in accordance with fitness function values.

Artificial bee colony algorithm (Benyoucef et al., 2015) is a swarm intelligence algorithm based on the honey bees forage activities. In artificial bee colony working bees search for food and share these records among the colony (Gonzalez-Castano et al., 2021). Jobless bees who attempt to determine a source of food by inspecting the working bees arbitrarily looking for a new source of food. In a short time, they interact and work with each other to achieve an optimal solution.

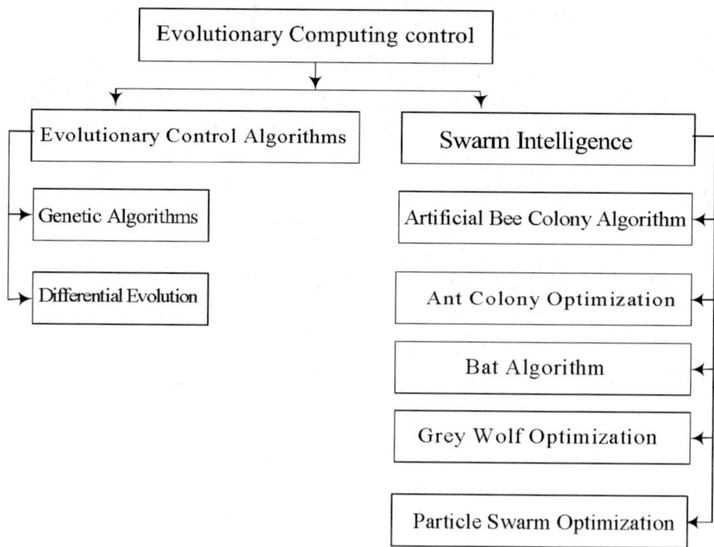

Figure 15. Evolutionary soft computing control.

Ant colony is another swarm intelligence optimization technique (Priyadarshi et al., 2019; Yaich et al., 2022) stimulated by foraging activities of the ants. With its attractive intensity, every ant in the population is attracted to other ants. The ants move from the lower to the higher potency zone on the basis of attractive force. This Attractive force is evaluated after each iteration and ants travel towards the optimum solution.

The Bat Algorithm (Kaced et al., 2017; Zhao et al., 2021) is driven by the activities of bats while looking for food. Main concept behind BAT algorithm (Kaced et al., 2017) is

- As bats use echolocation to sense the distance and can identify the distance of prey, surrounding obstacles & the variation in the existing prey or food in the hunt route.
- All Bats fly randomly at a particular position with particular velocity, loudness, fixed frequency varying wavelength in search of prey. On the proximity of prey's, they change/regulate their frequency/wavelength and rate of pulse emission.
- As the distance from the food decreases, the intensity of the bats changes to a least set rate.

Bat Algorithm begins its quest using an initial random population.

Another swarm intelligence technique is grey wolf optimization motivated by hierarchy and hunting nature of grey wolves (Mohanty et al., 2016). They live a severe social predominant chain of importance. In their society alpha wolves are the pioneer and chief, beta and delta wolves help the alpha in dynamic (Debnath et al., 2020). Omega wolves which are the lower part of the pyramid and ought to submit to the wide range of various predominant wolves. Major steps in their hunting are

a) Getting, observing and moving toward the prey
b) Looking for, encompassing, and upsetting the prey
c) Assaulting the prey

Particle Swarm Optimization is propelled by the common conduct of the fish rushes and winged creatures (Sundareswaran et al., 2015; Figueiredo and Nayana, 2021). In this technique, a particle addresses the every entity in the swarm having position and velocity vector. These particles chase a similar conduct: Each particle changes its state to best location of the swarm whose worth is the most nearest to the objective. The algorithm instates with a gathering of arbitrary arrangements and attempts to locate the best an incentive by refreshing every iteration.

Comparison between Various MPPT Techniques

Table 2 shows the comparison between all above discussed MPPT techniques on the basis of various parameters generally considered by an in individual while selecting a technique for his/her application.

Table 2. Comparative study of different MPPT techniques

Parameters / Techniques	Complexity	Efficiency	Speed of Convergence	Control policy	Cost	True MPPT
P&O	Low	Medium	Varies	Online	Medium	Yes
INC	Medium	Medium	Varies	Online	Medium	Yes
FOCV	Low	Low	Medium	Offline	Low	No
FSCC	Medium	Low	Medium	Offline	Low	No
FLC	High	Very High	Very Fast	Probabilistic	High	Yes
ANN	High	Very High	Very Fast	Probabilistic	High	Yes

Conclusion

Efficient use of solar energy not only saves conventional sources of energy but also brings financial profit in electric power generation and utilization. Many problems related to electrical energy can be solved by energy efficient solutions provided by solar PV systems. An effective energy efficient system is one which provides maximum power in its whole series of energy adaptation starting from generation to distribution. Photovoltaic system can be made more efficient by adopting many MPPT control schemes. MPPT control system is basically a power conditioner interface connected in between PV module and load. It is driven by a control unit to extort utmost power from in module in changing atmospheric conditions. Several MPPT techniques are available to perform this task. Various online, offline and soft computing methods are elaborated in the chapter with their advantages and limitations. Attempt is made to demonstrate each technique through flow charts. Moreover, a comparative table is made on certain useful parameters of each technique, which can help the one in choosing particular technique for his/her specific PV system requirement.

References

Abouadane, H., Fakkar, A., Sera, D., Lashab, A., Spataru, S. and Kerekes, T. (2020). Multiple-Power-Sample Based P&O MPPT for Fast-Changing Irradiance Conditions for a Simple Implementation. *IEEE journal of photovoltaics,* 10(5):1481-1488.

Ahmad, J. (2010). A fractional open circuit voltage based maximum power point tracker for photovoltaic arrays. In *Proceedings of the 2nd International Conference on Software Technology and Engineering,* San Juan, PR, USA, 1:V1-247–V1-250.

Alajmi, B., Ahmed, K., Finney, S. and Williams, B. (2011). Fuzzy-logic-control approach of a modified hill-climbing method for maximum power point in microgrid standalone photovoltaic system. IEEE Trans. *Power Electron,* 26(4): 1022–1030.

Alajmi, B., Ahmed, K., Finney, S., Williams, B. and Wayne, B. (2011). A maximum power point tracking technique for partially shaded photovoltaic systems in microgrids. *IEEE Transaction on Industrial Electronics,* 60(4): 1596–1606.

Ali, A., Almutairi, K., Padmanaban, S.K., Tirth, V., Algarni, S., Irshad, K., Islam, S., Zahir, M.H., Shafiullah, M. and Malik, M.Z. (2020). Investigation of MPPT Techniques under Uniform and Non-Uniform Solar Irradiation Condition-A Retrospection. *IEEE ACCESS,* 8:127368-127392.

Alik, R., Jusoh, A. and Sutikno, T. (2015). A Review on Perturb and Observe Maximum Power Point Tracking in Photovoltaic System. *Telkomnika,* 13(3):, 745~751.

Baimel, D., Shkoury, R., Elbaz, L., Tapuchi, S. and Baimel, N. (2016). Novel optimized method for maximum power point tracking in PV systems using Fractional Open Circuit Voltage technique. In *Proceedings of the 2016 International Symposium on Power Electronics, Electrical Drives, Automation and Motion,* Anacapri, Italy: 889–894.

Baimel, D., Tapuchi, S., Levron, Y. and Belikov, J. (2019). Improved Fractional Open Circuit Voltage MPPT Methods for PV Systems. *Electronics,* 8 (3).

Bandaru, R.K., Muraleedharan, C., and Kumar, M.V.P. (2019). Modelling and dynamic simulation of solar-thermal energy conversion in an unconventional solar thermal water pump. *Renewable Energy,* 134: 292–305.

Bastidas-Rodriguez, J.D., Franco, E., Petrone, G., Ramos-Paja, C.A. and Spagnuolo, G. (2014). Maximum power point tracking architectures for photovoltaic systems in mismatching conditions: a review. *IET power electronics,* 7(6): 1396-1413.

Bekker, B. and Beukes, H.J. (2004). Finding an optimal PV panel maximum power point tracking method. *7th Africon conference in Africa, 1 and 2: Technology innovation* Book Series: IEEE Africon: 1125-1129.

Benlahbib, B., Bouarroudj, N., Mekhilef, S., Abdeldjalil, D., Abdelkrim, T., Bouchafaa, F., and lakhdari, A. (2020). Experimental investigation of power management and control of a PV/wind/fuel cell/battery hybrid energy system microgrid. *International Journal of Hydrogen Energy,* 45 (53): 29110-29122.

Benyoucef, A.S., Chouder, A., Kara, K., Silvestre, S. and Sahed, O.A. (2015). Artificial bee colony based algorithm for maximum power point tracking (MPPT) for PV systems operating under partial shaded conditions. *Applied Soft Computing,* 32: 38-48.

Bhatnagar, A.P. and Nema, B.R.K. (2013). Conventional and global maximum power point tracking techniques in photovoltaic applications: A review. *Journal of renewable and sustainable energy,* 5(3): 032701.

Bingöl, O. and Özkaya, B. (2019). A Comprehensive Overview of Soft Computing Based MPPT Techniques for Partial Shading Conditions in PV Systems, *Journal of Engineering Sciences and Design,* 7(4): 926-939.

Bouchafaa, F. and Boucherit, M.S. (2010). Modelling and Simulation of a gird connected PV generation system with MPPT fuzzy logic control. In: *Proceedings of 7th international multi-conference on systems on signals and devices.*

Chamanpira, M., Ghahremani, M., Dadfar, S., Khaksar, M., Rezvani, A. and Wakil, K. (2019). *A novel MPPT technique to increase accuracy in photovoltaic systems under variable atmospheric conditions using Fuzzy Gain scheduling.* Energy Sources Part A-Recovery Utilization and Environmental Effects.

Cheng, P., Peng, B., Liu, Y., Cheng, Y. and Huang, J. (2015). Optimization of a fuzzy-logic-control-based MPPT algorithm using the particle swarm optimization technique. *Energies,* 8(6): 5338–5360.

Chian-Song, C. (2010). T–S fuzzy maximum power point tracking control of solar power generation systems. *IEEE Trans Energy Convers*: 1123–32.

Christopher, I.W. and Ramesh, R. (2013). Comparative Study of P&O and InC MPPT Algorithms. *American Journal of Engineering Research,* 02(12):402-408.

Debnath, D., Soren, N., Pandey, A.D. and Barbhuiya, N.H. (2020). Improved Grey Wolf assists MPPT Approach for Solar Photovoltaic System under Partially Shaded and Gradually Atmospheric Changing Condition. *International Energy Journal* 20 (1), 87 – 100.

Elgendy, M.A., Zahawi, B. and Atkinson, D.J. (2013). Assessment of the Incremental Conductance Maximum Power Point Tracking Algorithm. *IEEE transactions on sustainable energy,* 4(1): 108-117.

Femia, N., Granozio, D., Petrone, G., Spagnuolo, G. and Vitelli, M. (2007). Predictive & Adaptive MPPT Perturb and Observe Method. *IEEE transactions on aerospace and electronic systems,* 43(3): 934-950.

Femia, N., Petrone, G., Spagnuolo, G. and Vitelli, M. (2005). Optimization of Perturb and Observe Maximum Power Point Tracking Method. *IEEE transactions on power electronics,* 20(4): 963-973.

Fernandez, R., Ortiz, C., Chacartegui, R., Valverde, J.M., and Becerra, J.A. (2019). Dispatchability of solar photovoltaics from thermochemical energy storage. *Energy Conversion and Management,* 191: 237–246.

Firdaus, A.A., Yunardi, R.T., Agustin, E.I., Nahdliyah, S.D.N. and Nugroho, T.A. (2020). An improved control for MPPT based on FL-PSO to minimize oscillation in photovoltaic system. *International Journal of Power Electronics and Drive System,* 11(2):1082-1087.

Freeman, D. (2010). *Introduction to Photovoltaic Systems Maximum Power Point Tracking.* Texas Instruments Incorporated, SLVA446.

Frezzetti, Manfredi, S. and Suardi, A. (2014). Adaptive FOCV-based control scheme to improve the MPP tracking performance: An experimental validation. *Proceedings of the 19th World Congress The International Federation of Automatic Control Cape Town, South Africa,* 47(3): 4967–4971.

Gonzalez-Castano, C., Restrepo, C., Kouro, S. and Rodriguez, J. (2021). MPPT Algorithm Based on Artificial Bee Colony for PV System. *IEEE Access,* 9: 43121-43133.

Gupta, A., Chauhan, Y.K. and Pachauri, R.K. (2016). A comparative investigation of maximum power point tracking methods for solar PV system. *Solar Energy,* 136: 236-253.

Gupta, A., Kumar, P., Pachauri, R.K. and Chauhan, Y.K. (2014). Performance analysis of neural network and fuzzy logic based MPPT techniques for solar PV systems. *2014 6th IEEE Power India International Conference* (PIICON), Delhi, India.

Hannan, M.A.S. (2009). Maximum Power Point Tracking in Grid Connected PV system using a novel fuzzy logic controller. In: *Proceedings of IEEE student conference on research and development.*

Hart, G.W., Branz, H.M. and Cox III, C.H. (1984). Experimental tests of open loop maximum- power-point tracking techniques. *Solar Cells,* 13:185–95.

Hiyama, T., Kouzuma, S. and Imakubo, T. (1995). Identification of optimal operating point of PV modules using neural network for real time maximum power tracking control. *IEEE Transactions on Energy Conversion,* 10(2): 360-367.

Huang, Y.P. (2014). A rapid maximum power measurement system for high-concentration photovoltaic modules using the fractional open-circuit voltage technique and controllable electronic load. *IEEE Journal of photovoltaics,* 4(6):1610-1617.

Hussein, A., Hirasawa, K., Hu, J. and Murata, J. (2002). The dynamic performance of photovoltaic supplied dc motor fed from DC–DC converter and controlled by neural networks. In proceedings of the 2002 international joint conference on neural networks, Book Series: *IEEE International Joint Conference on Neural Networks* (IJCNN): 607-612.

Ishaque, K., Salam, Z., Amjad, M. and Mekhilef, S. (2012). An Improved Particle Swarm Optimization (PSO) Based MPPT for PV with Reduced Steady-State Oscillation. *IEEE Transactions on Power Electronics,* 27(8): 3627-3638.

Islama, M.T., Hudaa, N., Abdullahb, A.B., and Saidur, R. (2018). A comprehensive review of state-of-the-art concentrating solar power (CSP) technologies: Current status and research trends. *Renewable and Sustainable Energy Reviews,* 91: 987–1018.

Kaced, K., Larbes, C., Ramzan, N., Bounabi, M. and Dahmane, Z.E. (2017). Bat algorithm based maximum power point tracking for photovoltaic system under partial shading conditions. *Solar Energy,* 158: 490-503.

Kumar, P., Pachauri, R.K. and Chuahan, Y.K. (2015). Duty Ratio Control Schemes of DC-DC Boost Converter Integrated with Solar PV System. *2015 International Conference on Energy Economics and Environment (ICEEE):*1-6.

Lian, K.L., Jhang, J.H. and Tian, I.S. (2014). A Maximum Power Point Tracking Method Based on Perturb-and-Observe Combined With Particle Swarm Optimization. *IEEE Journal of Photovoltaics,* 4(2): 626-633.

Lopez-Lapena, O. and Penella, M.T. (2012). Low-power FOCV MPPT controller with automatic adjustment of the sample & hold. *Electronics letters,* 48(20).

Madaci, B., Chenni, R., Kurt, E., Hemsas, K.E. (2016). Design and control of a stand-alone hybrid power system. *International Journal of Hydrogen Energy,* 41(29): 12485-12496.

Masoum, M.A.S. and Sarvi, M. (2005). Design, simulation and implementation of a fuzzybased maximum power point tracker under variable irradiance and temperature condition. *Iran Journal Science Technology.*

Masoum, M.A.S., Dehbonei, H. and Fuchs, E.F. (2002). Theoretical and experimental analyses of photovoltaic systems with voltage and current-based maximum power-point tracking. *IEEE Transactions on Energy Conversion,* 17(4): 514–22.

Mellit, A. and Kalogirou, S.A. (2014). MPPT-based artificial intelligence techniques for photovoltaic systems and its implementation into field programmable gate array chips: Review of current status and future perspectives. *Energy,* 70: 1-21.

Mirza, A.F., Mansoor, M., Ling, Q., Khan, M.I., and Aldossary, O.M. (2020). Advanced Variable Step Size Incremental Conductance MPPT for a Standalone PV System Utilizing a GA-Tuned PID Controller. *Energies,* 13(16).

Mohanty, S., Subudhi, B. and Ray, P.K. (2016). A New MPPT Design Using Grey Wolf Optimization Technique for Photovoltaic System under Partial Shading Conditions. *IEEE Transactions on Sustainable Energy,* 7(1): 181-188.

Nkambule, M.S., Hasan, A.N., Hong, A.J., Geem, Z.W. (2020). Comprehensive Evaluation of Machine Learning MPPT Algorithms for a PV System under Different Weather Conditions. *Journal of Electrical Engineering & Technology.*

Noguchi, T., Togashi, S. and Nakamoto, R. (2002). Short-current pulse- based maximum-power-point tracking method for multiple photovoltaic-and-converter module system. *IEEE Transactions on Industrial Electronics,* 49(1): 217-223.

Noureddine, B., Djamel, B. and Boudjema, F. (2015). A hybrid fuzzy fractional order PID sliding-mode controller design using PSO algorithm for interconnected nonlinear systems. *CEAI,* 17(1): 41–51.

Pachauri, R.K., and Chauhan, Y.K. (2014). Fuzzy logic controlled MPPT assisted PV-FC power generation for motor driven water pumping system. *2014 IEEE Students' Conference on Electrical, Electronics and Computer Science,* Bhopal, India.

Priyadarshi, N., Ramachandaramurthy, V., Padmanaban, S., and Azam, F. (2019). An Ant Colony Optimized MPPT for Standalone Hybrid PV-Wind Power System with Single Cuk Converter. *Energies,* 12(1):167.

Ro, K. and Rahman, S. (1998). Two-loop controller for maximizing performance of a grid-connected photovoltaic-fuel cell hybrid power plant. *IEEE Transactions on Energy Conversion,* 13(3): 276-281.

Salam, Z., Ahmed, J. and Merugu, B.S. (2013). The application of soft computing methods for MPPT of PV system: a technological and status review. *Applied energy,* 107:135-148.

Sharma, N., and Puri, V. (2020). Solar Energy Fundamental Methodologies and its Economics: A Review. *IETE Journal of Research.*

Sharma, N., Bakhsh, F.I. and Mehta, S. (2018). Green independence. *International Journal of Scientific research in Computer Science, Engineering and Information Technology, National Conference on "Recent Advances in Computer Science and IT,* 4: 2456–3307.

Simoes, M.G., Franceschetti, N.N. and Friedhofer, M. (1998). A fuzzy logic based photovoltaic peak power tracking control. In *Proceeding of Indust Electron IEEE Int Symp,* 1:300–5.

Solanki, B.V., Bhattacharya, K. and Ca˜nizares, C.A. (2017). A sustainable energy management system for isolated microgrids. *IEEE Transactions on Sustainable Energy,* 8(4):1507–1517.

Subudhi, B. and Pradhan, R. (2013). A Comparative Study on Maximum Power Point Tracking Techniques for Photovoltaic Power Systems. *IEEE transactions on sustainable energy,* 4(1):89-98.

Sundareswaran, K., Vigneshkumar, V. and Palani, S. (2015). Application of a combined particle swarm optimization and perturb and observe method for MPPT in PV systems under partial shading conditions. *Renewable Energy,* 75: 308-317.

Tina, G.M., and Grasso, A.D. (2014). Remote monitoring system for stand-alone photovoltaic power plants: The case study of a PV-powered outdoor refrigerator. *Energy Conversion and Management,* 78: 862–871.

Wang, S.C., Pai, H.Y., Chen, G.J., and Liu, Y.H. (2020). A Fast and Efficient Maximum Power Tracking Combining Simplified State Estimation With Adaptive Perturb and Observe. *IEEE ACCESS,* 8: 155319-155328.

Wellawatta, T.R. and Choi, S.J. (2019). Regulated Incremental Conductance (r-INC) MPPT Algorithm for Photovoltaic Systems. *Journal of Power Electronics,* 19(6): 1544-1553.

Yaich, M., Dhieb, Y., Bouzguenda, M., and Ghariani, M. (2022). Metaheuristic Optimization Algorithm of MPPT Controller for PV system application. *E3S Web of Conferences* 336, 00036.

Zakzouk, N.E., Abdelsalam, A.K., Helal, A.A., and Williams, B.W. (2013). Modified variable step incremental-conductance MPPT technique for photovoltaic system. *39th annual conference of the IEEE industrial electronics society* (IECON 2013) Book Series: IEEE Industrial Electronics Society: 1741-1748.

Zakzouk, N.E., Elsaharty, M.A., Salam, A.K.A., Helal, A.A., and Williams, B.W. (2016). Improved performance low-cost incremental conductance PV MPPT technique. *IET renewable power generation, 10*(4): 561-574.

Zeitoury, J., Lalau, N., Gordon, J.M., Katz, E.A., Flamant, G., Dollet, A., and Vossier, A. (2018). Assessing high-temperature photovoltaic performance for solar hybrid power plants. *Solar Energy Materials and Solar Cells,* 182:61–67.

Zhang, L., Bai, Y.F. and Al-Amoudi, A. (2002). GA-RBF neural network based maximum power point tracking for grid-connected photovoltaic systems. In *Proceedings of international conference on power electronics, machines and drives,* Book Series: IEE conference publications, 487: 18–23.

Zhao, L., Jiang, M., Dadfar, S. Ibrahim, A., Aboelsaud, R. and Jamali, F. (2021). Grid-Tied PV-BES system based on modified bat algorithm-FLC MPPT technique under uniform conditions. *Neural Computing & Applications,* 33, 14929–14943.

Chapter 6

Fuzzy Logic-Based Maximum Power for Grid Connected PV Systems

Omveer Singh* **and Mukul Singh**
Department of Electrical Engineering, Gautam Buddha University, Gr. Noida, India

Abstract

The chapter aims to learn about MPPT and effectively implement the algorithms in code by using the Simulink/Simscape paradigm. To get maximum output, the authors have used the maximum power point tracking (MPPT) technique while implementing the fuzzy logic and simulated the model in MATLAB/Simulink. The complete analysis of efficient MPP technique is achieved at the initial stage of development. Two different techniques of MPPT are implemented in the research by the authors. The results of MPPT using the conventional Perturb & Observe (P&O) and intelligence-based fuzzy logic (FL) techniques are compared and analyzed. It is observed that to have more dependable control, fuzzy logic-based MPP is preferred over the conventional technique.

Keywords: standalone PV system, maximum power point tracking, fuzzy logic algorithm, voltage source converter, Perturb and Observe, renewable energy sources

*Corresponding Author's Email: omveers@gmail.com.

In: Applied Artificial Intelligence (AI) to Green Power Technology
Editors: Yogesh Kumar Chauhan, Ranjan Kumar Behera and Asheesh K. Singh
ISBN: 979-8-88697-131-6
© 2022 Nova Science Publishers, Inc.

Introduction

The irradiance, temperature, and current pulled from photovoltaic cells determine the magnitude of power obtained through a PV system with lots of cells. To get maximum output, the authors have used the maximum power point tracking technique. MPPT is useful in applications including putting electricity into the grid, charging the batteries, and powering the electric motors. In certain cases, the load may require more energy than the PV system can provide. In this case, power conversion is used to increase solar power. There is a variety of methods for optimizing the power output of a PV system, ranging from basic voltage relationships to more sophisticated multi-sample analysis. The power conversion engineer must weigh the various choices based on the final use and irradiance dynamics. Solar, wind, geothermal heat and tides are examples of renewable energy sources. These resources are renewed naturally and are deemed limitless for almost all-practical applications. The crisis of global energy has re-energized the gradual growth of sources from many organizations all over the world. Besides the world's rapidly diminishing fossil fuel supplies and reserves, another key issue involves mass atmospheric pollution involved with the emission of disastrous pollutants. RES in contrast to its traditional equivalents is thought to be safer and cleaner for generating energy without polluting the environment.

A Variety of Renewable Energy Sources

Wind Power

Now to capture the power present in airflow wind turbines are used. The rated power of today's turbines ranges between 600 kW and 5 MW as the power output relies on the wind speed and it depends on the cube of wind speed, it rises fast as the available wind velocity rises. Recent technological advances have resulted in the development of oil wind turbines, due to their superior aerodynamic structure, which is significantly more effective.

Solar Power

John Herschel is the inventor of the solar energy concept and when he went to Africa, a solar thermal energy box to cook the food was made by him. There are two main ways to use solar energy. To begin with, the collected heat could be taken as solar thermal energy for room heating. Converting incoming solar light into electrical energy and which is beneficial energy, and is another option. All this has been possible only because of the convenience of solar cells.

Small Hydropower

Small hydropower facilities with a capacity of less than 10 MW are classified as renewable energy sources [1]. Turbines are used to convert the stored energy in the liquid that is in the reservoirs into electricity. The purpose of river hydropower is to exploit the kinetic energy of water without using dams.

Biomass

Photosynthesis is the process by which plants harness solar energy. These plants release stored energy when they are burned. In this way, biomass acts as a natural battery for storing solar energy and releasing it as needed.

Geothermal

The thermal energy that is generated and trapped between various layers of Earth is known as geothermal energy [2]. The gradient so formed results in continual heat conduction from the earth's core to the surface. This gradient is used for heating the water to make steam, which is used for empowering steam turbines. The primary drawback of geothermal energy includes its geographical position. It is mainly found near tectonic plate boundaries, although many efforts have been made to improve it.

Trends of RES around the Globe

Figure 1 shows the recent scenarios in the industrial economy which shows that it is leaning toward renewable energy. In comparison to conventional power cities, North America and Europe have adopted greater renewable energy capacity in the previous three years. The recent advances in solar photovoltaic technology, as well as continuous project incubation in countries such as Germany and Spain, have driven huge growth in solar PV industries, which are expected to overtake a new renewable energy source-based development scenario in the coming years. India aims to produce 20 MW of solar power by 2022.

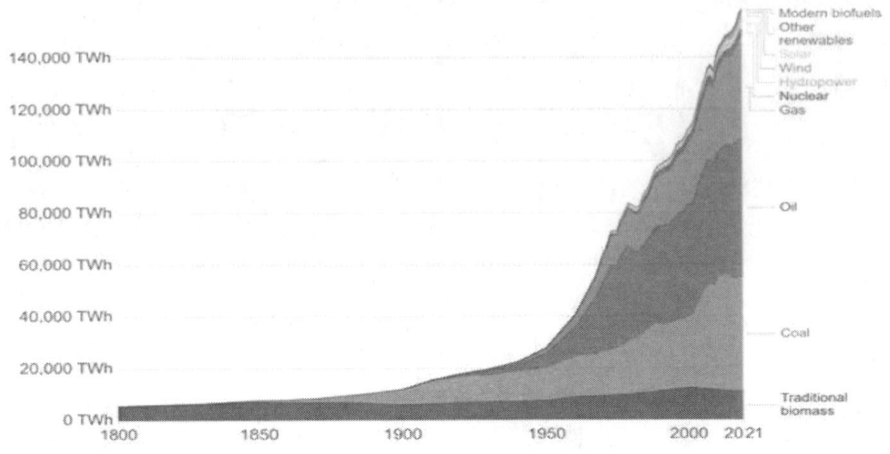

Figure 1. In the year 2020, increment in global energy consumption.

Solar Cell

Operating Principle

Photovoltaic panels are made up of solar cells. Although different materials are utilized, silicon is employed in the majority of them.

The photoelectric effect known as the capacity of many semiconductors to convert electromagnetic radiation directly into electricity is used in the solar cell. The charged particles created by incoming radiation are separated for generating an electrical current by solar cells. Figure 2 depicts a schematic of the p-n junction demonstrating the influence of a given electric field. On both sides, metallic contacts are placed to capture electrons and holes so that the current could pass. The contacts on the N-layer, which counteracts solar radiation, are several thin metal strips that enable light to flow into the solar cell, known as fingers. So far, the construction of solar cells is elaborated. Now its working principle is discussed further. Photons from the Sun fall on the solar cell. There are three different phases. Some photons reflect from the cell and the surface of the metal. Those who are not reflected get trapped in the substrate. Some photons, usually of very low energy, go unnoticed in the cell. Those that have an energy level over the band gap of silicon can only form an electron-hole pair. These pairs are present on either side of the p-n junction. The electric field propagates minority charges to the junction and causes them to spin in opposite directions to one side. This generates an electric current in the cell, which gets accumulated on both sides of metal contacts. Figure 2 depicts the current generated by light, which depends on irradiation. If irradiation is more than more, photons will come and create more electron-hole pairs so that solar cells generate more current.

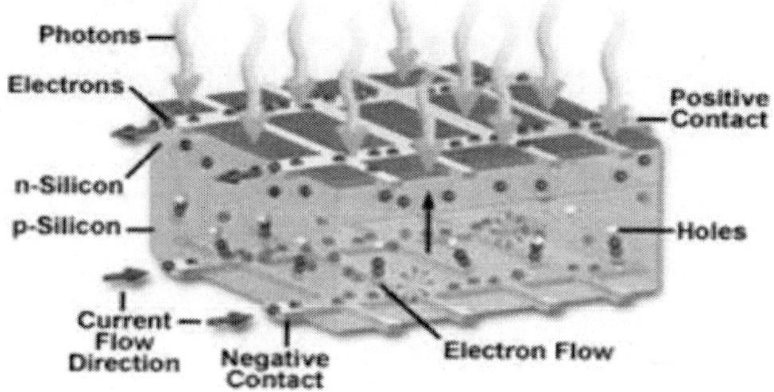

Figure 2. Solar cell.

The Need of Renewable Energy

Alternative energy sources include tidal waves, sunlight, rainfall, geothermal energy, and wind. These resources will never run out since they are supplied on demand. In contrast to these limitless natural resources, which are quickly diminishing the earth's storage capacity, the major source of energy today is fossil fuels. For a long time, people all over the world have been seeking for non-conventional energy sources to satisfy their vital energy demands. With the higher use of fossil fuels and exposure to the atmosphere, the adoption of clean and green mechanisms by countries all over the world has become a necessity. The sustainable and pollution-free use of green energies is what has drawn the current world towards itself. As a result, a significant amount of money is invested in harvesting these commodities. Authorities and organizations all over the world are implementing new schemes and regulations to increase the renewable strength utility in cutting-edge strength machines. Solar photovoltaic power, a well-known and well-connected generation for power technology within the renewable energy category, is presently exploding. People all around the world are becoming more aware of the significance and benefits of solar photovoltaic energy systems. However, the power of the sun alone is insufficient to meet the demand of customers. As a result, grid-related machine emerges as a feasible alternative. The demand for strength is pushed into the grid-related machines of the consumers. So subsequent viable answer emerges as the grid-related machine. In grid-related machines, the requirement of strength is being crammed with the aid of using a limitless supply of strength referred to as grid and sun photovoltaic strength. Thus, the evolution of grid-related photovoltaic machines evolved.

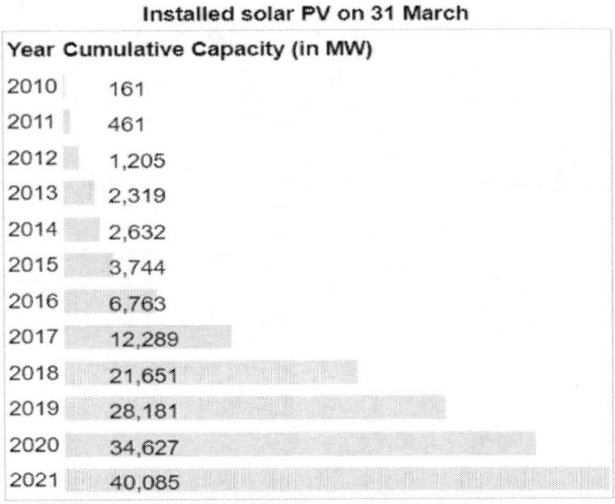

Installed solar PV on 31 March

Year	Cumulative Capacity (in MW)
2010	161
2011	461
2012	1,205
2013	2,319
2014	2,632
2015	3,744
2016	6,763
2017	12,289
2018	21,651
2019	28,181
2020	34,627
2021	40,085

Figure 3. Advancement of solar power in India.

By 2022, the Indian authorities have set a lofty purpose of producing 175 GW of pollution-free electricity. According to the Ministry of New and Renewable Energy (MNRE), India's anticipated renewable strength capability is 900 GW, with 20 GW coming from small hydro, 102 GW from wind strength (eighty-meter mast height), 25 GW from biomass, and 750 GW from sun strength (primarily based totally on three percentage wasteland). Because of the preceding facts, alternative energy has emerged as a significant supply of inexperienced energy among renewable energy, which can create history for the country's development and propel it to new heights within the Indian power industry.

Since conventional energy supplies are short, the increasing power demand of everyday life is cannot be fulfilled solely by them. Renewable energy sources have become more common as a complement to traditional systems to satisfy the energy demands. The primary vitality factors used in this respect are inexhaustible sources such as sun-oriented vitality and wind vitality. The consistent utilization of non-renewable energy sources has radically influenced nature, exhausting the biosphere and causing unnatural weather changes. Collecting solar vitality is conceivable in light of its accessibility. Sun-based vitality can be a stand-alone producing unit or a lattice-connected producing unit, depending on the availability of a surrounding framework. Because rural regions have limited access to the infrastructure, renewable energy sources are used to their full potential. Another advantage of solar energy is its universality, which allows it to utilize it wherever required. The present energy problem will solve if power could be effectively generated from solar radiation. In recent years, power transmission methods are substantially simplified. Power electronics and material technology advancements have assisted specialists in designing extremely compact but efficient systems to fulfill the load requirements. High power density is the sole drawback of these devices. R_S is the resistance provided by solar cell connections and bulk semiconductor material. R_{SH} (shunt resistance's) origin is more difficult to explain. It deals with the p–n junction's non-ideal character and the existence of impurities at the cell's borders that give a short-circuit path across the junction [3]. In an ideal situation, R_S would be zero and R_{SH} would be limitless. The impact of the shunt resistance is avoided due to the model's simplicity, resulting in R_{SH} being infinite. However, perfect condition is impossible to achieve, therefore manufacturers strive to reduce the impact of both resistances to enhance their goods.

The Mathematical Equation for MPP

The V_{OC} (open circuit voltage) and the I_{SC} (short circuit current) are two key aspects of the voltage-current characteristics. The power generated is zero at both locations. When the cell's output current is zero, i.e., I=0 and R_{SH} (shunt resistance) is discarded, V_{OC} is estimated from (1). Equation 1 is used to express it. As stated in the formula, at V = 0, I_{SC} current is nearly equivalent to I_L (generated light's current).

$$V_{OC} \approx (nkT/q) \ln(I_L/I_O)+1 \tag{1}$$

Solar cells generate the most power at peak in the V-I characteristic where the multiplication of V and I is greatest. The peak is known as point of maximum power and it is solitary, as seen in Figure 4, which depicts the prior points.

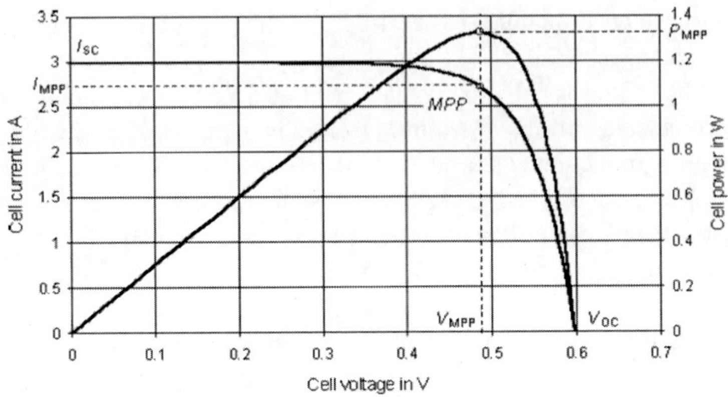

Figure 4. Important spots in a solar panel's output characteristics.

Literature Review

PV Noppadol Khaehintung et al. [4] proposed an adaptable self-organizing fuzzy logic controller (SOFLC) for solar traffic light equipment (SPTLE) with an MPPT integration system on an inexpensive micro-controller. It includes a boost converter for SPTLE with excellent performance. The PIC16F876A RISC microcontroller the duty ratio of the DC-DC boost converter. Villalva et. al. [5] in their article reported the analysis of current ecological and climatic circumstances, notably sun-oriented irradiance, temperature, and wind speed. At the same time, the plant provides timely information such as PV module temperature, power, current, and voltage. As the temperature of the environment rises, the power decreases. Under halfway hiding conditions, another Fuzzy Logic-based approximation is implemented to substantially decrease the power drop. H. Fujita et al. [6] propose a solitary pulse width modulation technique to demonstrate how a three-stage sun-powered force conditioner works. In each PWM-transporter cycle, the proposed pulse width modulation switches on or off one of the three-stage legs. In comparison to a traditional three-stage PWM approach, the presented method allows to decrease the usual exchanging recurrence and reduce the exchanging power misfortune to 1/3 or less. S.S. Raghuwanshi et. al. [3] propose and suggest a photovoltaic (PV) interface model with associated controllers. In the MATLAB/SIMULINK computer, the reproduction model is developed. A PV exhibit, dc-dc converter, PWM dependent voltage source inverter, and channel are the basic components of a single-stage framework linked PV framework.

The most intense force point following (MPPT) measures is used to ensure the efficiency of the PV framework. The reproduction studies demonstrate that irradiation of the sun increases the PV exhibit yield capacity, and active and reactive power (PQ). Pratima Mohanty et al. [7] in their article introduced several MPPT strategies and suggested some new ones. The author's chapter compares the efficiency and study of MPPT techniques used in solar PV generation systems under various temperature and irradiation level pressures, as well as in the direction of successful solar cell usage that can overcome conventional MPPT techniques. In his article, they discussed P&O, INC, constant voltage topologies, and control schemes. C. S. Chin et al. [8] introduced a fuzzy-based MPPT in a solar panel. In MATLAB/SIMULINK, the PV system is developed and evaluated. The fuzzy-based MPPT has higher performance and produces more electricity from solar panels, according to simulation findings.

Prof. Dr. Ilhami Colak and colleagues used MATLAB Simulink real-time analysis software to simulate three independent solar farms that each generates 15 kW of power [1]. MPPT algorithms are used in the P&O structure for each converter to convert energy. S. Yuvarajan et al. [9] presented a quick and accurate (MPPT) method for a PV panel based on the PV panel's OC voltage and SC current. The method was created using mathematical equations that described the PV panel's nonlinear V-I characteristics. Traditional approaches, such as the P&O, have recently been improved using artificial intelligence (AI) techniques. These algorithms cannot only improve energy efficiency, but they can help to mitigate the issues that conventional algorithms like P&O. A quick and reliable MPPT method for PV modules is suggested depending on the OC voltage and SC current of photovoltaic modules. The technique is based on mathematical formulae that determine the VI PV panel's non-linear properties. The MPPT algorithm may be used in a variety of circumstances, including those with different degrees of insulation, temperature, and wear. In the light intensity and temperature displayed by commercial photovoltaic modules, the overall difference in peak performance is less than 1.5%. The full version of the MPPT algorithm has been released. This algorithm has proven to be faster than other MPPT algorithms. The MPPT algorithm works in a variety of situations, including various levels of insulation, temperature, and deterioration. The technique is tested in MATLAB, and the results produced were found to be extremely near to the theoretically throughout a temperature range and light levels. For the light levels and temperatures that a PV panel would face, the highest variance in maximum power was less than 1.5 percent. This MPPT algorithm's entire derivation was provided. The technique is shown to be quicker than existing MPPT algorithms such as perturbation and observation and highly precise than approximation methods that rely on voltage linearity. Prof. Dr. Ilhami Colak and colleagues used MATLAB Simulink real-time analysis software to simulate three independent solar farms that each generates 16 kW of power [2]. MPPT algorithms are used in every converter's P&O topology to convert energy. S. G. Tesfahunegn and colleagues developed a novel solar/battery charge controller which integrates MPPT and overvoltage controls into a single control function [10]. To construct the used dual-loop control arrangement. The proposed controller has been shown for excellent transient responsiveness with relatively minor voltage overshoot. Yuncong Jiang et al. provide an analog MPPT controller for a PV system that uses load side current to maximize the solar panel's output power [11]. The suggested circuit is less expensive and smaller than the conventional MPPT controller circuit.

Simulation results confirmed the suggested MPPT controller's tracking performance. Ma T. et al. [12] described a novel hypothetical model that struck a good balance between precision and simplicity. In MATLAB, a model for determining sunlight-based photovoltaic (PV) module parameters is developed, and the model was then fitted to exploratory I-V trademark bends of a PV module/string/cluster. For the model, only a few sources of data are needed, which can be found on the creator datasheet. The appearance of a PV module/string/cluster at any sun-based irradiance and module cell temperature is recreated with this recently developed model. Arash S. et al. presented a unique MPPT scheme aimed primarily at applications related to fixed loads and charging of battery-based applications. This is accomplished by output current maximization [13]. This method has several advantages, including a typical current controller and circuit scheme independence. This results in excellent energy conversion efficiency in low-cost low-power applications. For simulation reasons, a novel hybrid PV model is presented. Finally, simulation results are presented, demonstrating the algorithm's validity. Solar photovoltaic (PV) systems have been a focus of study for a decade, to increase the performance of solar PV modules because the I-V curve of a solar PV module is non-linear, and some approach is required to monitor the voltage and current at peak point on I-V curve that corresponds to MPPT. As a result, techniques like "Maximum Power Point Tracking" (MPPT) are frequently used. Although several MPPT approaches are evolved in recent years, P&O is applied because of its efficient algorithm, ease of implementation, and low cost. MPPT approach is utilized in most commercial applications [12]. Moreover, in case of rapidly changing irradiance conditions, this method is slow to monitor MPP and vibrates around it. Ali F Murtaza et al. address the P&O technique's problematic behavior by providing a modern MPPT hybrid technique that combines two typical methods, namely P&O and FOCV techniques to get better of the P&O technique's inherent flaws. In comparison to P&O, the suggested MPPT approach is more reliable in monitoring the MPP though under quickly varying irradiance. When subjected to changes in irradiance, the approach is confirmed using Simulink simulated outputs indicating an enhancement in reaching the MPP.

Simulation Models and Blocks

PV Modelling

When electrons move from N-side to the P-side of the junction and if there is an obstacle in the way, it is called series resistance. Parallel resistance is generated due to leakage current. An electrical field is created within a solar PV cell when irradiance strikes the cell's surface. This process divides negative and positive charges in an absorbing medium as shown in Figure 5 (joining p-type and n-type). These charges are able for generating the current that is utilized in an external circuit when they are exposed to an electric field. The intensity of the incoming radiation determines the produced current. The more the electrons that are released from the surface, the more the current is generated, the higher will be the light intensity [14].

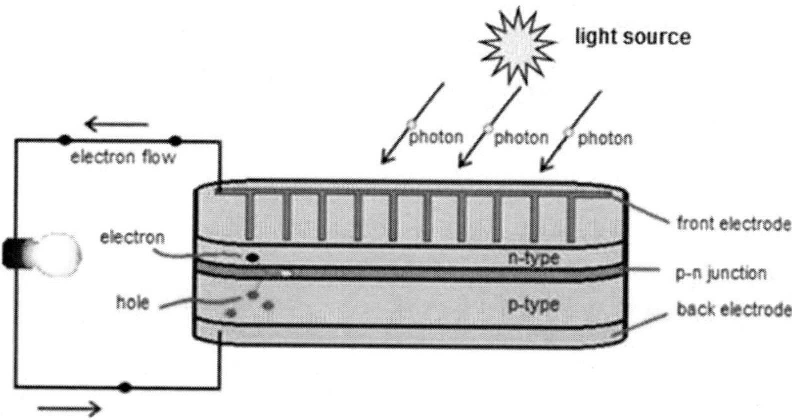

Figure 5. A cross-section prototype of a typical solar cell.

PV cell model is the most critical factor that influences the simulation's accuracy. While modeling a PV cell, we check the value of the real cell at each environmental condition. We connect a current source in parallel with a diode to model an ideal solar cell. The shunt resistance and series resistance are shown in Figure 6 [15, 16].

I_{pv} denotes the current source of the photocell, the series resistance of the photocell is denoted by R_s, and the shunt resistance of the photocell is denoted by R_{sh}. The value of shunt resistance is mostly higher as compared to series resistance so to make the calculations easier, authors neglect the value of series resistance sometimes.

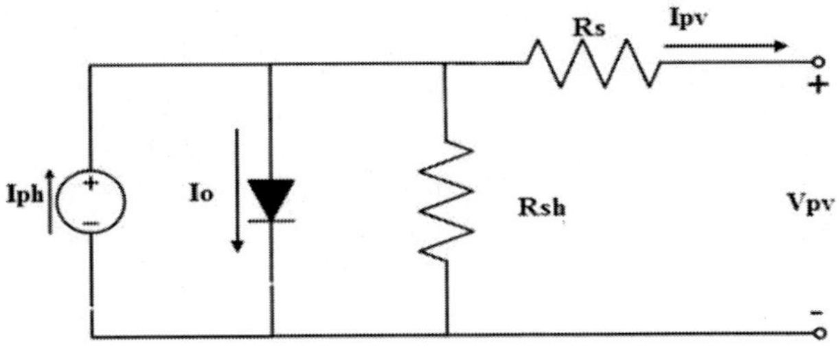

Figure 6. Photovoltaic cell equivalent circuit.

The following equations provide the mathematical model of PV that researchers utilized to clarify our PV array.

$$I = N_pI_{ph} - N_pI_s * [\exp\{q(V+R_sI)/N_sAkT\}-1] \tag{2}$$

$$I_{ph} = I_{ph}(T_{ref})[1+a(T-T_{ref})] \tag{3}$$

$$I_{ph}(T_{ref}) = I_{sc}(T_{ref})*(E/E_o) \tag{4}$$

$$a = [\{I_{sc}(T_2)-I_{sc}(T_1)\}/I_{sc}(T_1)] * (1/T_2-T_1) \tag{5}$$

$$I_d = I_s * [\exp \{q(V+R_sI)/AkT\}-1] \tag{6}$$

$$I_s = I_{so} * (T/T_{ref})3/A * \exp [(-E_g/A_k)*\{(1/T)-(1/T_{ref})\}] \tag{7}$$

$$I_{so}=I_{sc} (T_{ref})/[\exp\{qV_{oc}T_{ref}/AkT_{ref}\}-1] \tag{8}$$

$$R_s = (-dV/dI_{Voc})-(1/X_v) \tag{9}$$

$$X_v = I_{so} * \{q/AkT_{ref}\}\exp\{qVoc(T_{ref})/AkT_{ref}\} \tag{10}$$

Table 1. Parameters of green solar India 37W (AT-37) PV module

Parameters and values	
P_{max}(W)	36.920
V_{mp}(V)	17.910
I_{mp}(A)	2.064
I_{sc}(A)	2.231
V_{oc}(V)	21.423
N_s	36
N_p	1

Photovoltaic Cell Simulink Model in MATLAB

The formula for the conversion of temperature (°C to °F) is as follows:

$T_{rk} = 273 + 25$ (ref. temp.)

$T_{ak} = 273 + T_{op}$ (operating Temp.)

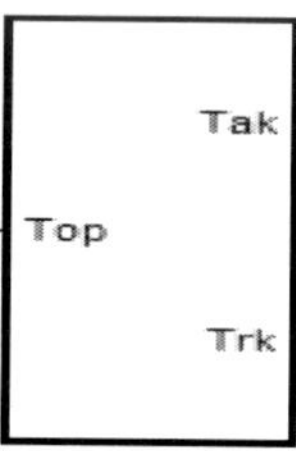

Figure 7. Block for temperature conversion.

Figure 7 shows the block for temperature conversion. When at least one of these modules is damaged or hidden, this diode's capability is to direct the current. Another diode, known as a blocking diode, is connected to every string to eliminate the current stream from switching and to secure the modules.

The basic equation of a single solar cell is:

$$I = I_{PV} - I_O[\exp(qV/akT)-1] \tag{11}$$

"I" stands for the net current of a single solar cell, "I_{pv}" stands for the total current created by solar radiation, "I_0" stands for the reverse saturation current or diode leakage current, and "a" stands for the ideal diode ratio, that is, the minimum diode ratio required. Adjust the values to compare the solar cell's PN junction with the observed value; "T" is the diode's Kelvin temperature, electronic charge is denoted by "q," and Boltzmann constant is denoted by "k." The properties of real solar cells are not well described by Equation (1). The model may better adjust to empirical situations by adding series and parallel resistors and monitoring the terminal voltage [17].

$$I = I_L - I_O[\exp\{(V+R_{ser}I)/V_t a\}-1] - [(V+R_{ser}I)/R_{per}] \quad (12)$$

"R_{ser}" represents the overall series confrontation of resistance for all connected solar cells, "R_{per}" represents the entire parallel confrontation of resistance for all connected solar cells, and "V_t" is the terminal voltage, which can be calculated as "$V_t = (kT / q) = T / 11600$ "and voltage at the combination terminal of the solar cell is denoted by "V." Contact resistance between the connection terminal and the solar cell generates "R_{ser}," and "R_{per}" is generated by the leakage current of the PN junction. (2) The photovoltaic system are can be modified by arranging the series or parallel connection of solar cells.

Effect of Load Mismatching

We can see from the P-V characteristics of PV modules that there will be only one point at which power is highest with V_{MPP} as the associated voltage and I_{MPP} as the corresponding current. The maximum power is transmitted to the load if the load line crosses this point. This load resistance value is calculated as shown in Figure 8 [18, 19].

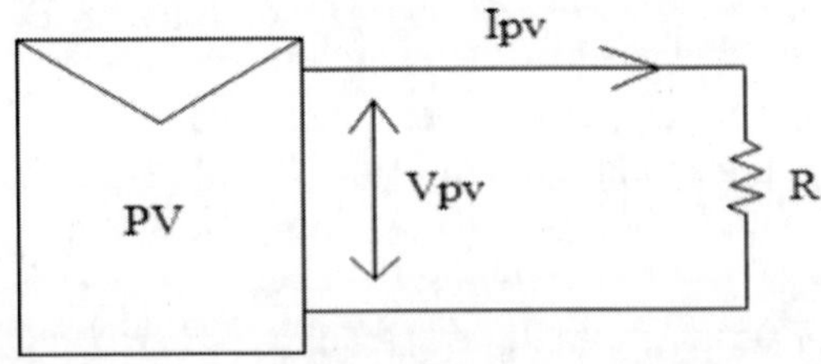

Figure 8. PV interfacing to load.

The capacity of a photovoltaic depends on its size and type of solar cell. The load on the output side has a big impact on the output power. In case of load mismatch, the delivered power is cannot be maximized. The difference between the internal source resistance and the internal load resistance on the output side is known as load mismatch.

If the source's equivalent resistance is the same as the load resistance, then maximum power is delivered, this is the MPPT, R_{eq} is known as characteristic impedance and can be easily determined from the manufacturer's datasheet. The V_{MPP} and I_{MPP} values are in the table so we can determine the characteristic impedance. In this case, the R_{MPP} is 7.9.

Boost Converter

With an input range of 6–23 volts, a "boost converter" is to increase the constant output voltage to 24 volts. The reference value and the output voltage are compared, and a PWM waveform is created. DC to DC converter boosts the voltage from 12V to 24V. The duty cycle, D, is set to 50%. Using a microcontroller to build the boost converter for a photovoltaic is shown in the research chapter. The purpose of the converter is to convert SPP to a stable 24V output without the use of any storage devices such as batteries. A voltage-feedback approach is used to regulate this through an FPGA unit.

Figure 9 shows the circuit diagram of the boost converter. When the switch is closed, energy is taken and stored by the inductor. The inductor current grows slowly in this mode, although the charging and discharging of the inductor are linear for simplicity's sake. Because the diode prevents current to flow. The capacitor's discharge current maintains a constant load current.

Figure 10 shows the functioning of the boost converter in model 1. In mode 2, the diode is short-circuited since the switch is open. The inductor's stored energy is released via opposite signs, which charge the capacitor. Across the procedure, the load current stays unchanged. Figure 1 shows the waveform of a "boost converter."

As shown in Equation 13 the output and input voltage of "boost converters" rely on their duty cycle (D) value (1).

$$V_{out} = Vin/(1 - D) \tag{13}$$

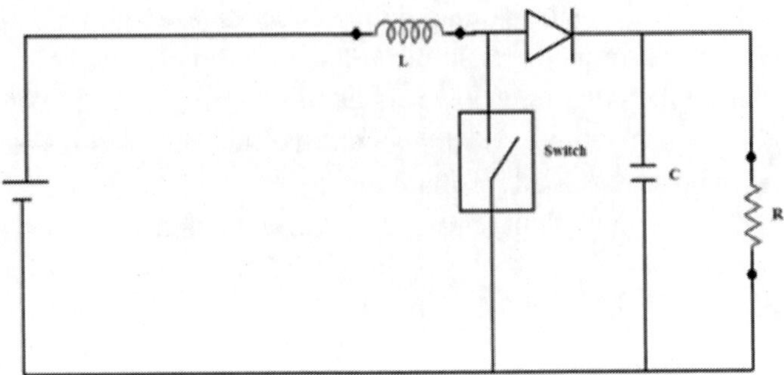

MODE 1: Functioning of the boost converter.

Figure 9. Circuit diagram of a boost converter.

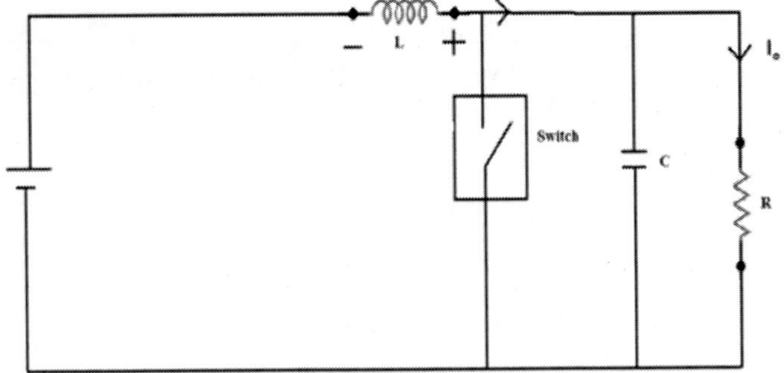

MODE 2: Functioning of the boost converter.

Figure 10. Functioning of the boost converter in model 1.

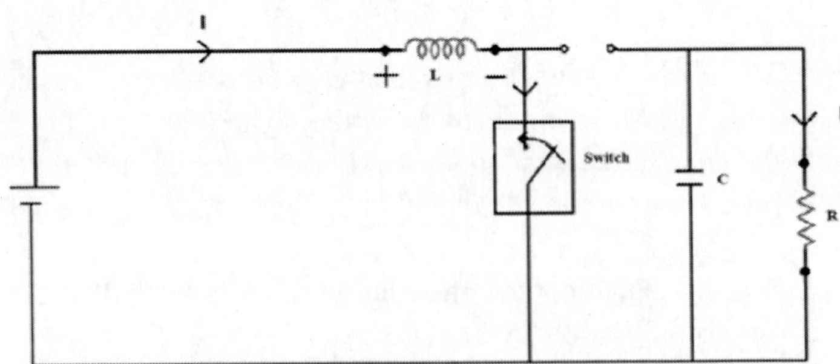

Figure 11. Mode 2 functioning of the "boost converter."

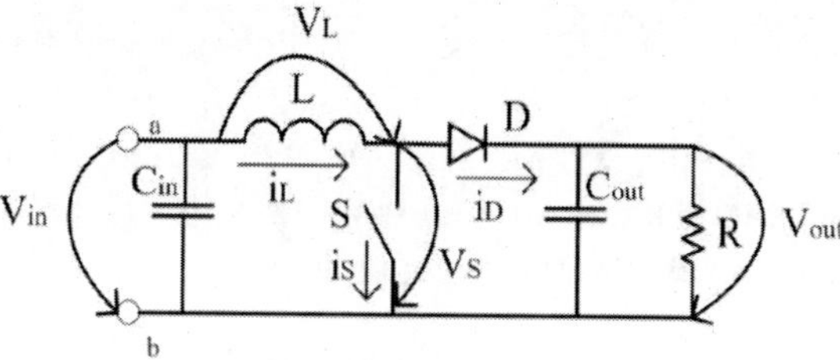

Figure 12. General boost converter with input and output capacitor filter.

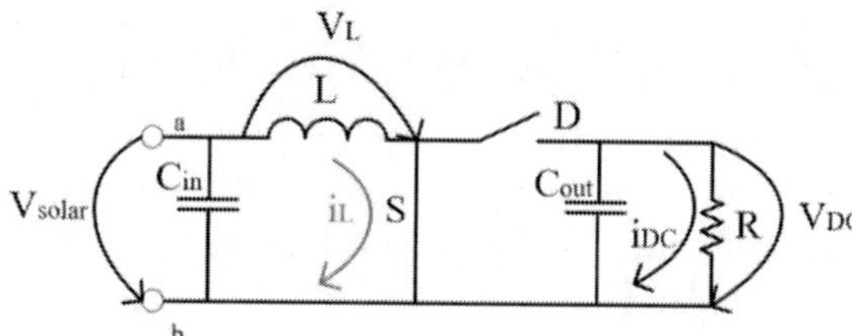

Figure 13. Mode 2 functioning of boost converter with input and output capacitor filter.

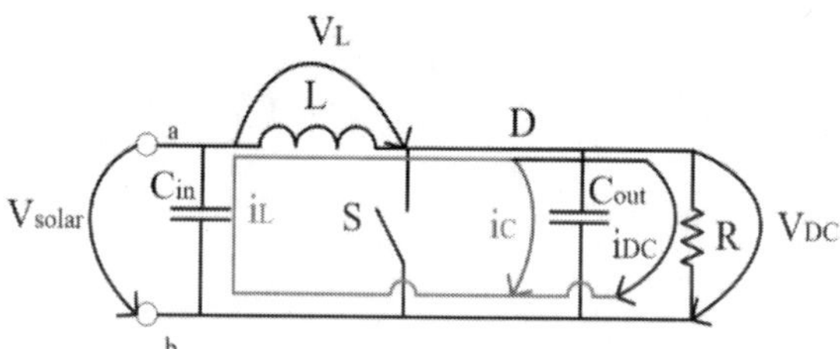

Figure 14. Mode 1 functioning of boost converter with input and output capacitor filter.

Figures 12, 13, and 14 show the working of the boost converter in mode 2. The voltage in this system output is higher than the voltage at system input. This is determined by the sum of the solar panels' voltages. Since the voltage at the output is high, the converter is named a boost converter because the coil switches polarity to avoid a rise in current. The diode, coil, capacitor, and switch, which is an IGBT, must all be able to support a peak voltage value of 1100 V and a peak current value of 13 A for this PV system. Equations (2)–(4) are used to calculate the input and output capacitor capacitances, as well as the coil inductance. Capacitors are adjusted for ripple voltage. (DV). the input voltage is 160V. The parameters are calculated for the duty cycle of 0.80, which is the allowed peak value described in Chart 1, and the voltage would be 800 V at the output. The diode selected is a pair of Vishay VS-15EVL06-

M3 Schottky diodes connected in series, which can withstand voltages over 1100V, the voltage drop of 1.96V. Infineon IGW60T120 was chosen as the IGBT. 1.90 V is the voltage drop. The 20 kHz switching frequency was used for these components (f). It is advised that the target value be 0.40, while any number between 0.30 and 0.60 is acceptable.

Procedure for Designing a Boost Converter

Load Requirement: The load used is a conventional 4x4 LED panel, which requires a current of 11- 16 mA per row of four LEDs on a line. The load requirement is 570 watts with a total current of 42 mA, requiring a voltage of 24 volts. Because a potential divider with a total resistance of 1100 is utilized, the total equivalent resistance becomes R_{eq} = (1100) (570) = 375. The parameters are tabulated in the following Table, along with the other factors that will be determined based on this load requirement.

Table 2. For the boost converter, these values are used

S. No.	Parameters	Value
1.	Inductor.	295μH
2.	MOSFET 1N5408.	IRFi 850
3.	Power Diode.	IN5407
4.	Input Capacitor.	480μF
5.	Output Capacitor.	350 μF
6.	Resistive Load.	55Ω, 60W

Maximum Power Point Tracking Algorithms

A Study on MPPT Techniques

The most frequently applied algorithms for obtaining MPP are P&O and Incremental Conductance or by using PID or PI (Proportional and Integrated) controller, as they work only regardless of the environment and are coded to obtain a particular value. They will not attempt to secure the MPP for the changing environmental conditions under these circumstances. Other AI-based methodologies offer even greater outcomes whether used alone or in conjunction with traditional MPPT algorithms. In the upcoming sections, both the simple P&O approach and the intelligence with fuzzy reasoning methodology will be studied in detail [20].

Algorithm for Fuzzy Logic

During the last decade, microcontrollers have made fuzzy logic control for MPPTs standards. As stated before, Fuzzy logic controllers offer multiple advantages, including the ability to handle specific inputs, not requiring precise mathematical models, and handling non-linearity. Fuzzification, Rule Basis Table Lookup, and Deffuzification are the three phases of fuzzy logic control. As presented in Figure 15, fuzzification includes a membership function, to convert a numerical input variable into a linguistic variable. For added accuracy, reference and seven fuzzy levels are employed. Figure 15 is mainly composed of a numerical variable's range of values. To provide more weight to particular fuzzy levels, the subscription function is often adjusted less symmetrically.

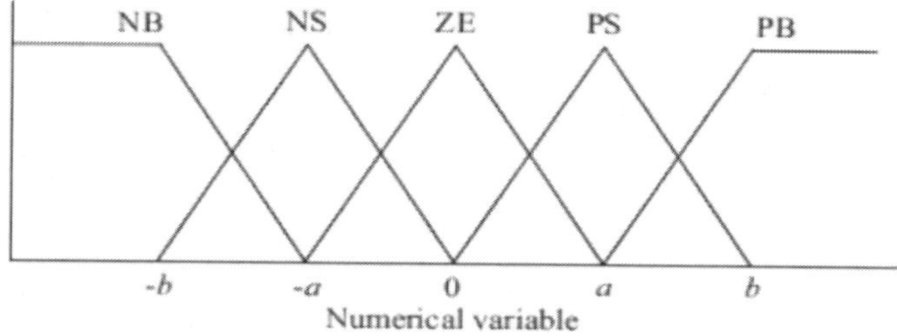

Figure 15. Fuzzy logic controller inputs and outputs have a membership function.

As previously indicated, fuzzy logic has just recently been developed for MPPT-based PV systems. It is utilized in conjunction with other AI-based methods like artificial neural networks or evolutionary algorithms to improve the efficiency of specific MPPT methods (such as P&O, incremental conductivity, or PID). Traditional MPPT algorithms that have been improved are more stable, and their design procedures do not require accurate model knowledge. The primary stages of a fuzzy logic controller are divided into three categories: fuzzification, inference rules, and diffusivity. Figure 16shows these steps as well as the controller's proposed plan. Fuzzification is the initial phase, which implies mixing crisp inputs with stored membership functions and assigning them to each input to create the fuzzy input. After the initial phase of fuzzification, to generate its input values, the controller will compare real-time inputs with stored membership function information.

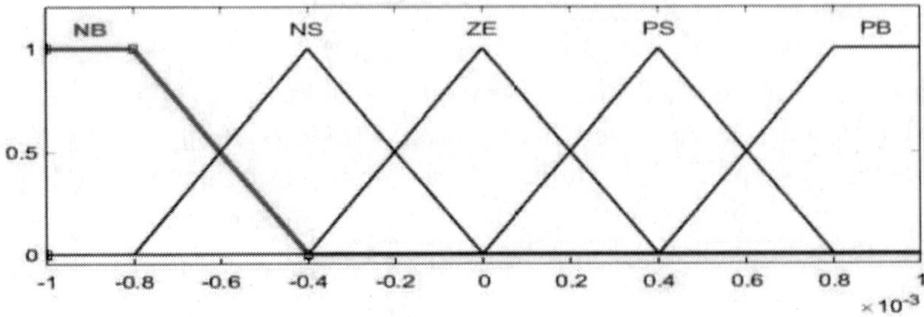

Figure 16. Fuzzy logic controller inputs and outputs have a membership function.

Fuzzification converts numerical input variables into linguistic variables using membership functions. To give additional weighting to particularly unclear levels, the membership function is altered less symmetrically.

Detailed Information of Perturb and Observe Algorithm

By applying a positive voltage disturbance, we can move closer to the MPP. MPP is located to the right of Point B. When a positive disturbance is applied, the value of P drops, necessitating a change in the direction of the disturbance to get MPP. In Figure 18, the P&O method is shown as a flowchart, as shown below.

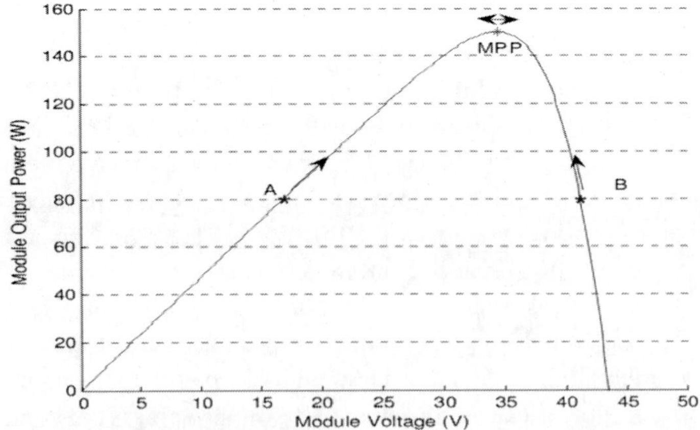

Figure 17. Characteristics of solar panels MPP and operating points A and B.

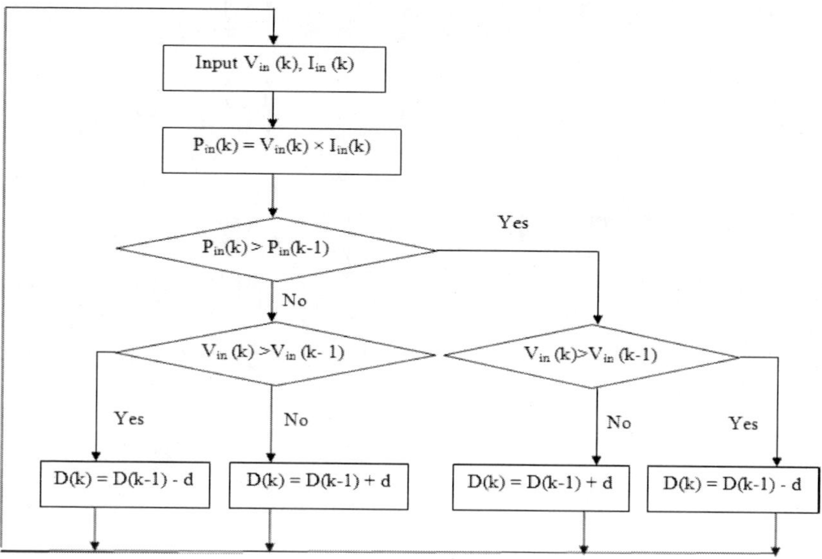

Figure 18. Flowchart of Perturb & Observe algorithm.

Implementation Method

Without batteries, inverters and solar PV panels are chosen to allow a conventional four-wire three-phase building. The values of inductors, capacitors, Schottky diodes, and IGBTs are determined using a three-phase inverter as a model, with a voltage ripple of 5 V, a value for the coefficient R is 0.40 and an operating frequency of 20 kHz. Fuzzy logic is used for PV system development. For that authors have used Green SolarIndia37W (AT-37) PV Module. This module generates 213.15 watts with every single PV module. For the system, five parallel with single series connection systems is used. The duty signal determines the duty ratio of PWM as it controls the power as well. When power and voltage are constantly varying, the derivate is zero and the point is at maximum. The goal is to optimize the power provided by PV systems under a variety of lighting conditions, while also improving efficiency and lowering the cost. The quality of output given to the load is used to measure the system's performance. Voltage, current, and frequency are major characteristics that determine the quality of delivered power in the case of electrical loads. By regulating the voltage parameter, the proposed system improves the power generation, which upgrades the power quality [21-24].

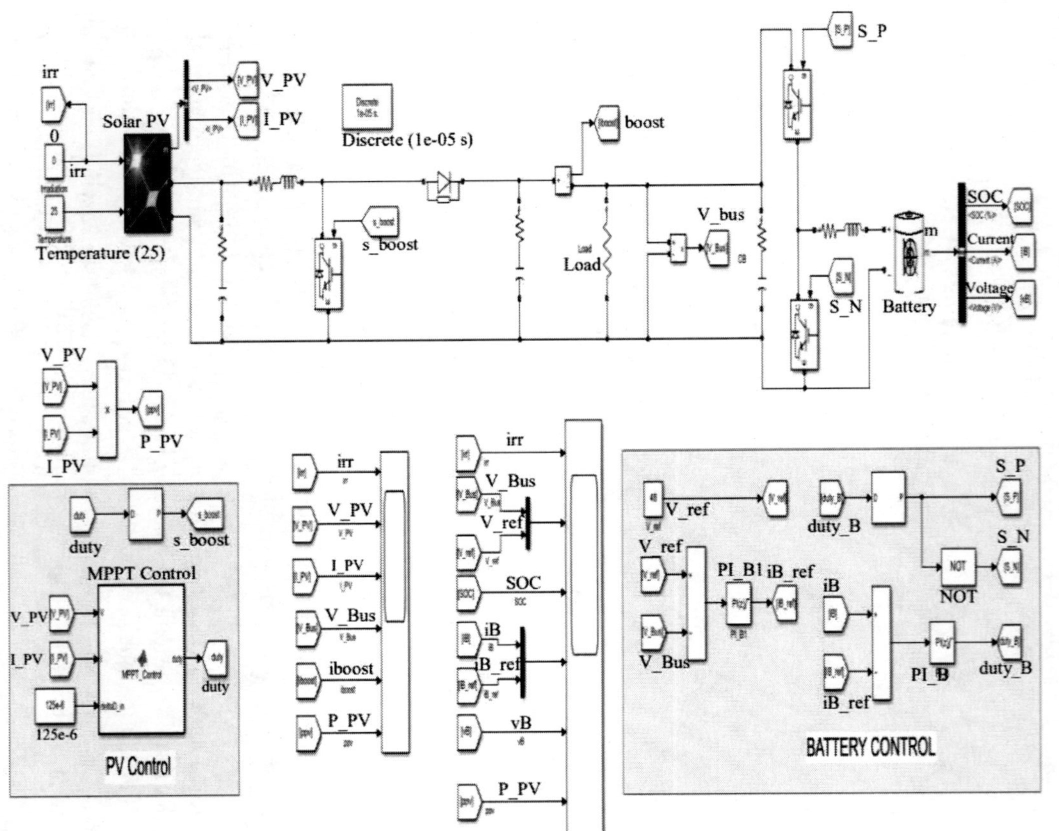

Figure 19. PV panel system with MPPT model (complete model).

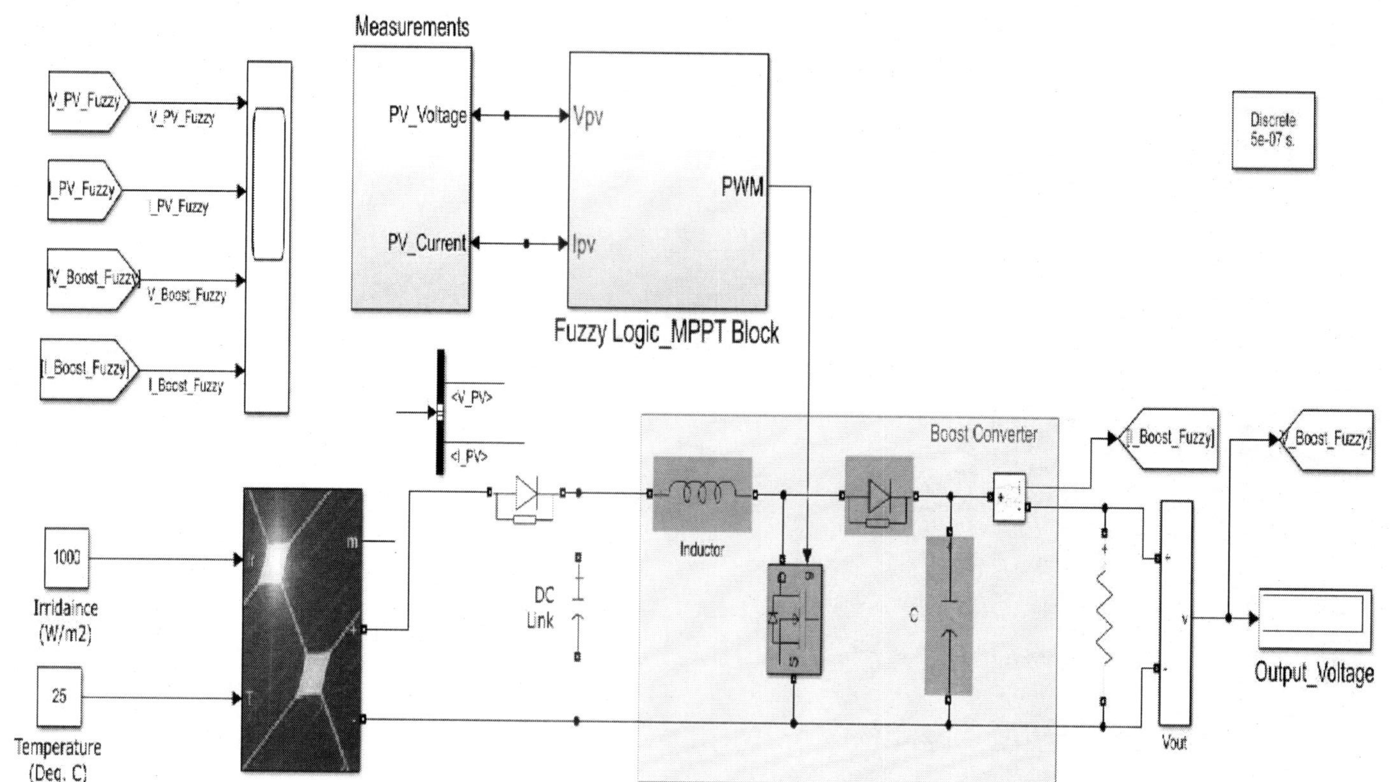

Figure 20. PV panel with fuzzy model (complete model).

Result and Discussion

Figure 21 shows the input irradiation graph value for the solar PV system. As we know that as the radiation of the sun increase, the generation of voltage in the Solar system also increases. This graph shows the radiation in increasing and decreasing steps just like sunrises and sunsets.

Figure 22 represents the PV system's input voltage. This graph shows the generated voltage through the PV system when the irradiation becomes 200 then the voltage generation starts and its increases as the irradiation increases.

Figure 23 shows the input current generated by the PV System. This graph shows the generated current by the PV system when the irradiation becomes 200 then the voltage generation starts and its increases as the irradiation increases.

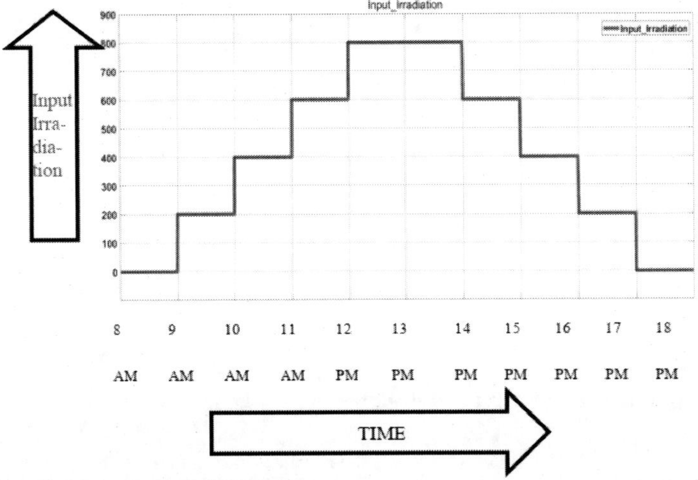

Figure 21. Input irradiation of PV system.

In Figure 24, the graph shows the voltage generated by the PV system at the load side when the irradiation becomes 200 then the voltage generation starts, and increases as the irradiation increases. The boost converter boosts its voltage.

In Figure 25, the graph shows the voltage generated by the PV system at the load side and generated side. When the irradiation becomes 200 then the voltage generation starts and it increases as the irradiation increases. The boost converter boosts its voltage. This graph shows the comparison between input and output voltage using P&O MPPT Techniques.

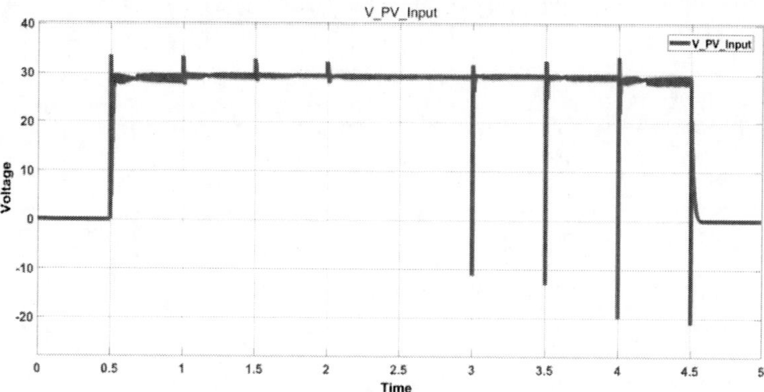

Figure 22. Input voltage generated by PV system.

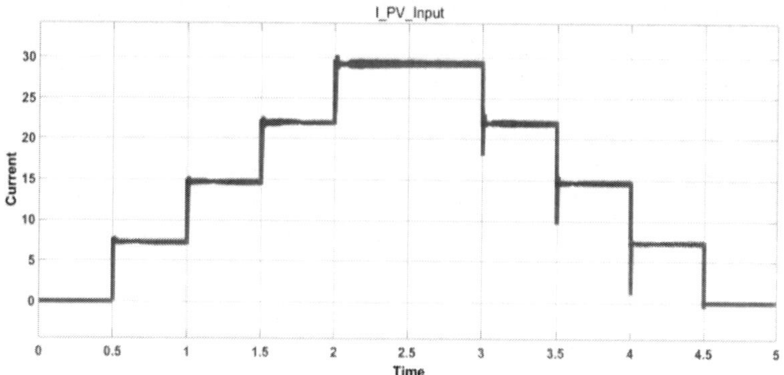

Figure 23. Input current generated by PV system.

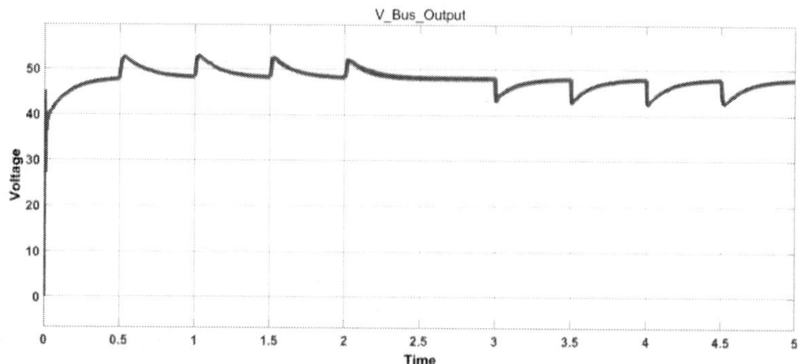

Figure 24. Output voltage at load side.

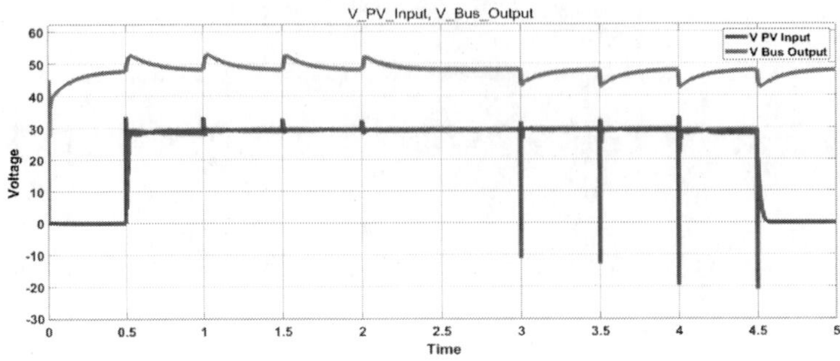

Figure 25. Input and output voltage comparison using P&O for MPPT.

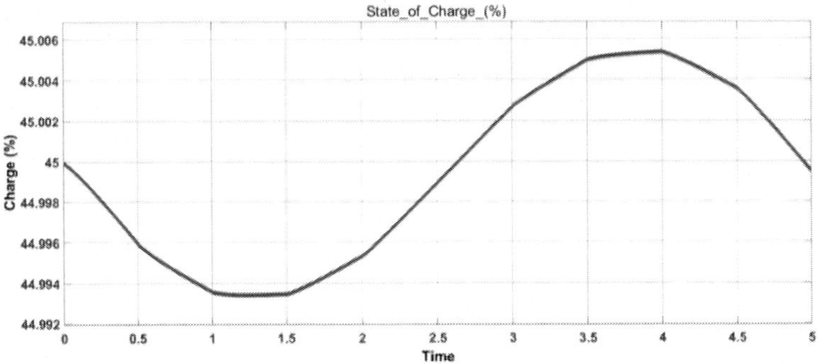

Figure 26. Charging status of battery.

Figure 26 shows the charging status of the Battery. As we see in the graph before time 2, it discharges because there is low irradiation after that it comes in charging mode because there we get good irradiation. This show that when there is low radiation, we can use power from the battery and when there is good irradiation then the battery will charge. The du signal determines the duty ratio of PWM and it controls the power by depending on MPP. MPPT algorithms determine the MPPs by searching the dP/dV=zero derivative. When power and voltage are constantly varying, the derivate is zero and the point is at maximum. Figure 21 represents the voltage output of a fuzzy logic-based PV system coming from the PV panels, 1000 w/m2 solar radiations were applied to the system and spread. The results for PV System with battery integration by using P&O MPPT techniques. Figure 21 represents the voltage output fuzzy-logic-based PV system coming from the PV panels, 1000 w/m2 solar radiations were applied to the system, spread over eight-time intervals.

Results for PV System with Battery Integration by Using Fuzzy Logic Algorithm MPPT Techniques

In Figure 27, the graph shows the voltage generated by the PV system at the load side and generated side. When, the irradiation becomes 200 then the voltage generation starts and increases as the irradiation increases. Its voltage is boosted by the boost converter. This graph shows the comparison between input and output voltage using the fuzzy logic algorithm MPPT Techniques.

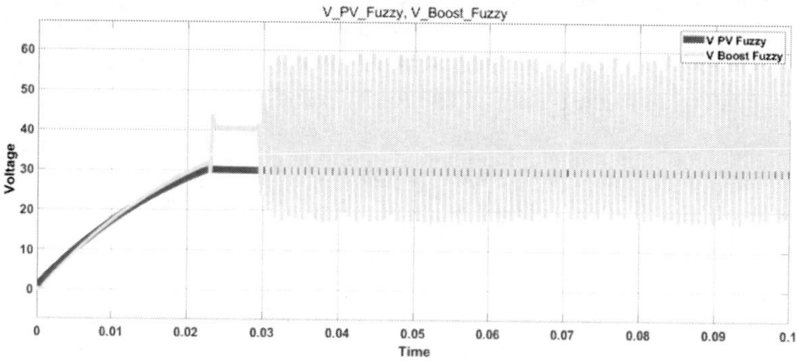

Figure 27. Input and output voltage comparison using fuzzy logic for MPPT.

Conclusion

Several "Maximum power point tracking" techniques are thoroughly examined. MPPT techniques are chosen on basis of implementation cost, the number of sensors required, and complexity. Tracker is operated by increasing or reducing the solar array voltage from time to time. MPPT is a key component of the system that performs well because it increases power generation from the PV system under a particular set of environmental and load conditions, maximizing array efficiency and lowering overall system costs. MPP alters with varied radiation and cell temperature, therefore it is important to keep track of it and build appropriate algorithms to keep the system running as efficiently as possible. The control approach for MPPT in a PV system is provided in the chapter for various radiation and temperature circumstances. The P&O method is the most widely used traditional method in this discipline and when compared to the fuzzy methodology, both techniques have been simulated and analyzed. This application is better suited to fuzzy technology, which provides more dependable control than P&O.

Future Scope

In this chapter, MATLAB Simulink models are designed as per standalone system with conventional and Fuzzy logic control methods with battery connection for power outage backup. For future scope, implementation of other soft computing methods are can be performed. Providing battery backup is a good option as it can be further integrated with the grid, which allows the overall system to track power in the two-way phenomenon. Model predictive control (MPC) process control can also be used and is considered an advanced method because it is mainly used to control a process in a set of constraints and power electronics. The dynamic model of the process assures the predictive controllers, linear empirical models are often achieved by the identification of the system. One of the advantages of MPC is that it allows adapting to the present time while keeping in mind the future time. Recently this method is trying to be implemented with electric vehicles but it can also work with solar PV systems while changing the solar irradiation. In this way, the power flow could be controlled by the MPC algorithm based on MPPT. Other than all these, for MPPT we can use better algorithm techniques, which may be based on the sensors or can be directly algorithm-based. Different methods include methods like incremental conductance methods or particle swarm optimization methods, etc.

References

[1] Chin, C. S. (2011). Fuzzy logic based MPPT for photovoltaic modules influenced by solar irradiation and cell temperature. *UK Sim 13th International Conference on Modelling and Simulation*.

[2] Tesfahunegn, S. G. Ulleberg, O. Undeland, T. M. and Vie, P. J. S. (2011). A simplified battery charge controller for safety and increased utilization in standalone PV applications. *International Conference on Clean Electrical Power (ICCEP)*.

[3] Fujita, H. A three-phase voltage-source solar power conditioner using a single-phase PWM control method. *Proc. of IEEE Energy Conversion Congress and Exposition*, pp. 3748-3754.

[4] Andrews, A. and Scaria, R. (2019). Three-phase single-stage solar PV integrated UPQC. 2nd *International Conference on Intelligent Computing, Instrumentation and Control Technologies (ICICICT)*, pp. 1130-1135.

[5] Khaehintung, N. and Sirisuk, P. (2007). Application of maximum power point tracker with self-organizing fuzzy logic controller for solar-powered traffic lights. *7th International Conference on Power Electronics and Drive Systems.*

[6] Villalva. Gradella, M. and Gazoli, J. R. (2009). Comprehensive approach to modeling and simulation of photovoltaic arrays. *IEEE Transactions on Power Electronics*, Volume 24 Issue 5, pp. 1198-1208.

[7] Raghuwanshi, S. S. and Gupta, K. (2015). A single-phase grid-connected photovoltaic system using MATLAB/Simulink. *IEEE International Conference on Computer, Communication and Control.*

[8] Mohanty, P. (2014). MATLAB based modeling to study the performance of different MPPT Techniques used for solar PV system under various operating conditions. Elsevier, *Renewal and Sustainable Energy Reviews 38,* pp. 581- 893.

[9] Colak, I. Kabalci, E. and Bal, G. (2011). Parallel DCAC conversion system based on separate solar farms with MPPT control. *8th International Conference on Power Electronics - ECCE Asia*, TheShillaJeju, Korea.

[10] Yuvarajan, S. and Shoeb, J. (2008). A fast and accurate Maximum Power Point Tracker for PV systems. *Twenty-Third Annual IEEE Applied Power Electronics Conference and Exposition.*

[11] Yang, B. Zhu, T. Wang, J. Shu, H. Yu, T. Zhang, X. Yao, W. and Sun, L. (2020). Comprehensive overview of maximum power point tracking algorithms of PV systems under partial shading condition. *Journal of Cleaner Production,* vol. 268, pp. 121983.

[12] Romo, D. Palomo, R. Rivero, R. L. Francisco, M. and Sanchez, S. (2020). Averaged current mode control for maximum power point tracking in high-gain photovoltaic applications. *Journal of Power Electronics,* vo. 20, iss. 6, pp 1650-1661.

[13] Lin, B. Wang, L. and Wu, Q. (2020). Maximum Power Point Scanning for PV systems under various partial shading conditions. *IEEE Transaction on Sustainable Energy,* vol. 11, iss. 4.

[14] Mao, M. Huang, H. Zhang, L. Chong, B. and Zhou. L. (2020). Maximum power exploitation for grid-connected PV system under fast- varying solar irradiation levels with modified SALP swarm algorithm. *Journal of Cleaner Production,* 122158.

[15] Sajjad, D. Samad, S. and Nakamura. H. (2019). Variable step size perturb and Dadfar observe MPPT controller by applying θ-modified krill herd algorithm-sliding mode controller to increase accuracy in photovoltaic system. *Journal of Cleaner Production,* vol. 271, 122243.

[16] Li, Y. Samad, S. Ahmed, F. W. Abdulkareem, S. S. Hao, S. and Rezvani, A. (2020). Analysis and enhancement of PV efficiency with hybrid MSFLA–FLC MPPT method under different environmental conditions. *Journal of Cleaner Production,* vol. 271, pp. 122195.

[17] Guo, B. Su, M. Sun, Y. Wang, H. Liu, B. and Zhang, X. (2020). Optimization design and control of single-stage single-phase PV inverters for MPPT Improvement. *IEEE Transactions on Power Electronics,* vol. 35(12), pp. 13000-13016.

[18] Khaehintung, N. and Sirisuk, P. (2007). Application of maximum power point tracker with self-organizing fuzzy logic controller for solar powered traffic lights. *7th International Conference on Power Electronics and Drive Systems (ICPEDS).*

[19] Sirisuk, P. (2018). A solar-powered battery charger with neural network maximum power point tracking implemented on a low-cost PIC-microcontroller. *7th International Conference on Power Electronics and Drives Systems (ICPEDS).*

[20] Jiang, Y. Hassan, A. Abdelkarem, E. and Orabi, M. (2012). Load current based analog MPPT controller for PV solar systems. *27th Annual IEEE Applied Power Electronics Conference and Exposition (APEC).*

[21] Singh, M. Singh, O. and Kumar, A. (2019). Renewable energy sources integration in micro-grid including load patterns. *3rd International Conference on Recent Developments in Control, Automation & Power Engineering (RDCAPE)*, pp.s 88-93, doi: 10.1109/RDCAPE47089.2019.8979036.

[22] Singh, M. Rana, V. Ansari, M. A. Dikshant, Saini, B. and Singh, P. (2018). Power quality enhancement to sensitive loads with PV based microgrid system. *International Conference on Sustainable Energy, Electronics, and Computing Systems (SEEMS),* pp. 1-6, doi: 10.1109/SEEMS.2018.8687334.

[23] Singh, M. and Singh, O. (2019). Phasor solution of a microgrid to accelerate simulation speed. *2nd International Conference on Advanced Computing and Software Engineering (ICACSE),* doi: 10.2139/ssrn.3351025.

[24] Singh, M. Ansari, M. A. Tripathi, P. and Wadhwani, A. (2018). VSC-HVDC transmission system and its dynamic stability analysis. *International Conference on Computational and Characterization Techniques in Engineering Sciences (CCTES),* doi: 10.1109/CCTES.2018.8674095.

Chapter 7

Different Reconfiguration Approaches for Photovoltaic Systems

Vijay Laxmi Mishra*, **Yogesh Kumar Chauhan and K. S. Verma**
Department of Electrical Engineering, Kamla Nehru Institute of Technology
Sultanpur, Uttar Pradesh, India

Abstract

Photovoltaic systems are considered to be more feasible among the existing renewable sources of energy that are capable of producing electricity when sunlight is incident on it. But an intricate issue hampering the solar photovoltaic array (SPVA) power generation is partial shading condition (PSC). Multiple power peaks are formed such as global maximum power point (GMPP) and local maximum power point (LMPP) on the output characteristics depicting loss in power occurred due to PSC. To mitigate such issues, SPVA reconfiguration strategies have been adopted. Several conventional approaches were developed that were quite not effective as compared to the existing advanced SPVA reconfiguration that involves many cross-ties but is capable of effective shade-dispersion improving the performance metrics under PSC. Paying heed to such issue, this book chapter focuses on various recent reconfigured models (Ken-ken (K-K), Arithmetic Sequence (AS), L-Shape (L-S)) along with conventional models Bridge-Linked (B-L), Series-Parallel (S-P), Total cross-tied (TCT)) under various shading, to understand the effect of performance parameters like GMPP, Fill Factor (FF), Efficiency (η), Execution ratio (ER), %Power Loss (%PL), and Mismatch Loss (ML) for a symmetrical model using

*Corresponding Author's Email: laxmi.2514@knit.ac.in.

In: Applied Artificial Intelligence (AI) to Green Power Technology
Editors: Yogesh Kumar Chauhan, Ranjan Kumar Behera and Asheesh K. Singh
ISBN: 979-8-88697-131-6
© 2022 Nova Science Publishers, Inc.

MATLAB. Thus, this book chapter illustrates how with the aid of unaltered electrical connections and physical relocations this advanced reconfigured model (L-S) outperforms conventional models (TCT).

Keywords: arithmetic sequence reconfiguration, Ken-ken reconfiguration, L-shape reconfiguration, solar photovoltaic module, total cross-tied

Introduction

Due to the increase in population, energy demand is rising to meet social and economic needs. Earlier human beings were dependent on fossil fuels for electricity generation. But due to its depletion entire population has switched toward renewable sources of energy. For the development of worldwide energy resources, non-conventional sources of energy have played a vital role [1]. Among the available renewable sources, solar energy is considered to be more sustainable as it is available in abundance, eco-friendly and cheaper [2]. The generation of electricity from the sun is governed by several constraints such as irradiation level and temperature. Solar energy generation is hampered when tall buildings, towers, passing clouds, dust, ice, and hailstorm get deposited on the solar panel leading to PSC as shown in Figure 1. To address such a problem, traditional models such as series (S), parallel (P), B-L, honey-comb (H-C), TCT, and triple-tied (TT); their hybrids such as S-P, TCT-BL, TCT-HC, BL-HC, etc. have been modeled in the literature [3-5]. These traditional models were not much effective in terms of performance parameter enhancement. Hence, advanced reconfigured models such as Sudoku (SDK) [6-7], dominance square (D-S) [8], magic square (M-S) [9-11], LoShu (LSh) [12], skyscraper (SS) [13-14], jig-saw (j-S) [15], odd-even (OE) [16], a novel prime number (NPN) [17], ken-ken (K-K) [18], ancient Chinese (AC) [19], etc. have been reported in the literature that is capable of minimizing power loss and enhancing GMPP without perturbation.

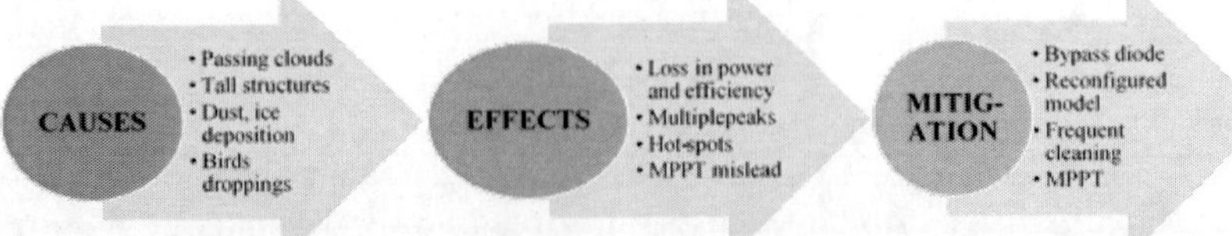

Figure 1. Illustration of PSC.

Multiple reconfigurations on different array sizes were observed. TCT, SP-TCT, BL-TCT, BL-HC, and M-S schemes of interconnection were applied on a 4x4 array of 170W. The M-S reconfiguration outruns in terms of %PL, FF, and GMPP. The simulation performed was on vertical, diagonal, and horizontal shading. The algorithm was able to track GMPP easily [20]. TCT, DS, and novel competence square (NCS) interconnection were simulated on a 9x9 array. NCS gave a high performance in terms of FF and %PL [21]. Latin square (LSq) interconnection was compared with TCT. The results showed lower ML in LS reconfiguration. The overall performance was observed to be reduced at lower irradiation levels [22]. Different types of hybrid interconnection were simulated as SP-TCT, BL-TCT, and, BL-HC and compared to TCT, SP, BL, HC, and D-S reconfiguration. The performance of D-S was better but the power loss was very high in all interconnections due to high dispersion losses [8]. An experimental setup was developed to test the performance of SS concerning the conventional schemes. The reduction in LMPP gave a better level of power production. The performance was reduced at low insolation levels [13]. An experimental setup was tested on ladder reconfiguration (L-R) for a 4x6 size array for six different shading patterns. A minimum power loss of 46% was observed for L-R as compared to SP, TCT, HC, and BL reconfiguration. The hardware implantation was costly for efficient reconfigurations [23]. Improved Sudoku (I-SDK) was simulated on a 9x9 array of 170W modules. In this work, GMPP was improved and losses were reduced but the hardware complexity was high [24]. Diffusion charge compensation (DCC) reconfiguration at hardware setup was developed and compared with modules with and without bypass connection. A total of twelve different shading patterns were observed. The performance ratio is observed to be 75% for DCC [25]. The performance of SP, BL, HC, TCT, R-TCT, and SDK was investigated on an experimental setup in terms of %PL, FF, and performance ratio (PR). SDK was observed to be better in terms of PR. The implementation was simpler but the hardware implementation was costly [7]. SS reconfiguration tested was on various shading conditions and compared with six other traditional and hybrid interconnections. Losses were reduced but the hardware implementation was complex [14]. OE scheme was applied for a 6x4 array of 200W module rating. It was simulated for different shading patterns. Performance measured for %PL, performance enhancement ratio (PER), and FF. %PL is observed to be reduced for OE as compared to TCT, SP-TCT, BL-TCT, and HC-TCT [16]. Reposition scheme was applied for 3x4, 5x8, and 7x8 arrays of 60W rating. The complexity was highly reduced in interconnections but the tracing of GMPP was tough. The M-S scheme was again tested in two different articles. Comparisons were made with SDK and TCT for 170W and 200W modules. A substantial reduction in power loss was observed [26-27]. TCT, K-K, OE, LS, and SDK reconfiguration were checked on simulation as well as experimental setup on the 4x4 size PV module and the K-K was observed to be the best scheme [18]. Similarly, K-K, SS, j-S, and OE reconfigurations were compared on a 4x4 size 200W module along with the conventional TCT schemes. J-S was observed to be best for various considered shaded schemes [15]. The 10 W module with a size of 4x4 was analyzed on simulation and experimental setup. Performance was measured in terms of %PL. The minimum %PL was observed to be 14% for the L-S [28]. Array sizes of 9x9 and 23x23 re simulated for different reconfiguration variants of SDK with DS and NPN reconfiguration. The NPN scheme gave a minimum mismatch loss of 4% [17].

This chapter is aimed at addressing various shading issues in terms of enhancing the performance parameters such as GMPP, Fill Factor (FF), Efficiency (Ƞ), Execution ratio (ER), %Power Loss (%PL), and Mismatch Loss (ML) with the help of conventional and advanced reconfigured models of SPVA for a symmetrical model built using MATLAB tool. Further, it is also depicted through a bar graph how efficient is the advanced reconfigured model (L-S) over the traditional models (TCT).

Mathematical Modelling of Solar Cell

The very basic step in understanding the function of a solar cell is its mathematical model. There are so many solar cell models such as the 1-diode model (1-D) [28], 2-diode model (2-D) [29], 3-diode model (3-D) [29], etc. studied in the literature under varying climatic factors to analyze its performance. Two major constraints influencing SPVA are irradiation level and temperature. Since the 1-D model is more effective than the 2-D model hence this chapter uses it for further investigation [28].

The 1-D solar cell model is shown in Figure 2 which comprises series and shunt resistances R_S and R_{SH}, a current source, and a shunt connected diode. Its output equations for voltage and current are formulated by equation (1) with the aid of nodal analysis.

$$I_{cell} = I_{ph} - I_D - I_{SH} \tag{1}$$

Equation (2) shows the relation between the output current and output voltage.

$$I_{cell} = I_{ph} - I_0 \left[\exp^{q\left[\frac{V_{cell} + I_{cell} R_S}{nkT}\right]} - 1 \right] - \frac{V_{cell} + I_{cell} R_S}{R_{SH}} \quad (2)$$

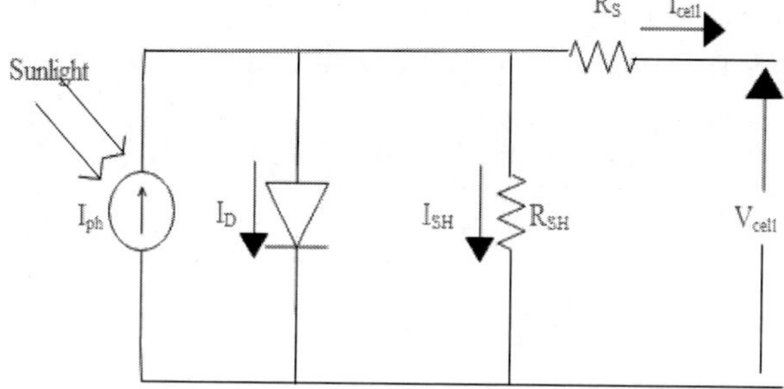

Figure 2. 1-D model of a solar cell.

Equation (2) violates under varying metrological factors due to the heating of solar cells and the formation of hot spots. It may also lead to a fire under major non-uniform conditions. Hence, to overcome this a bypass diode is placed on the sub-strings of the solar cell module [30]. Therefore, an alternative direction for the current passage is incorporated, preventing the module from malfunctioning. Table 1 shows values for a single 5W solar module used in this chapter.

A direct relation between irradiation level and short-circuit current (I_{sc}) can be observed in Figure 3 specifying varying irradiations under STC for a 5W solar module.

Figure 4 represents the plot of the current-voltage (I-V) and power-voltage (P-V) curve of the 5W solar module under varying temperatures showing clearly an inverse relationship between temperature and V_{OC}.

Various Modelling Topologies for Observing PSC Effects

Basic Connecting Topologies

This sub-section discusses a few conventional topologies as follows:

Series-Parallel (S-P)

S-P topology is achieved by connecting several solar modules in a string fashion in series and then these series-connected strings are connected in parallel. Thus, the required output voltage is obtained by series-connected strings and the output current is obtained through a parallel combination [1]. The merit of the S-P connection is that it is capable of achieving the required values of current and voltage precisely which is not accurately achieved in series and parallel connections when done individually. But, S-P is not capable of reducing ML to much extent and is less efficient under PSC. Figure 5 (a) shows 4 × 4 S-P connected SPVA, where 4 strings are linked together in a shunt to generate the required output current. The SPVA current (I_{SPVA}), voltage (V_{SPVA}), and power (P_{SPVA}) are represented by equations (3), (4), and (5).

$$I_{SPVA} = I_1 + I_2 + I_3 + I_4 = 4I_m \quad (3)$$

$$V_{SPVA} = V_1 + V_2 + V_3 + V_4 = 4V_m \quad (4)$$

$$P_{SPVA} = V_{SPVA} I_{SPVA} = 16 V_m I_m \quad (5)$$

Table 1. Data for a single 5W solar module

Parameters	Values
Maximum Power, P_{Max}	5W
Maximum Voltage, V_{Max}	9.62V
Maximum Current, I_{Max}	0.52A
Open Circuit Voltage, V_{OC}	11.25V
Short Circuit Current, I_{SC}	0.55A
Number of series-connected cells, N_{Se}	18
Irradiation level and temperature under STC	1000 W/m^2 and 25°C
Irradiation level under PSC	450 W/m^2
Dimension (L×W×H)	(255×185×17) mm

Bridge-Linked (B-L)

B-L connection involves solar module connections in a bridge fashion as shown in Figure 5 (b). The two solar modules are initially linked in series and then finally linked in parallel orientation. It comprises fewer connections but ML and η are still less compared to S-P [3]. The SPVA current (I_{SPVA}), voltage (V_{SPVA}), and power (P_{SPVA}) are represented by equations (6), (7), and (8).

$$I_{SPVA} = 4I_m \quad (6)$$

$$V_{SPVA} = 4V_m \quad (7)$$

$$P_{SPVA} = V_{SPVA} I_{SPVA} = 16 V_m I_m \quad (8)$$

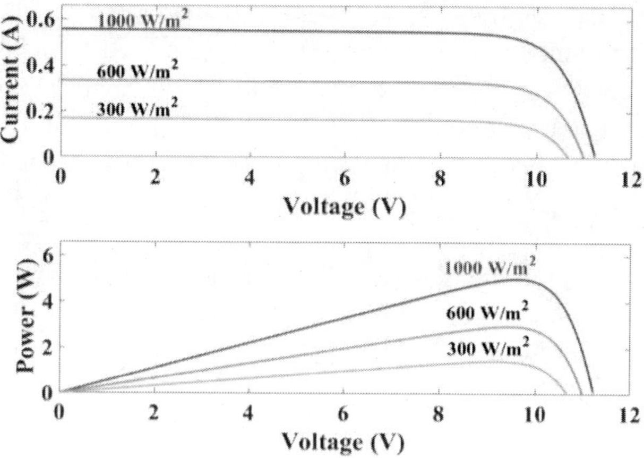

Figure 3. I-V and P-V plot under specified irradiances at 250C for a 5W module.

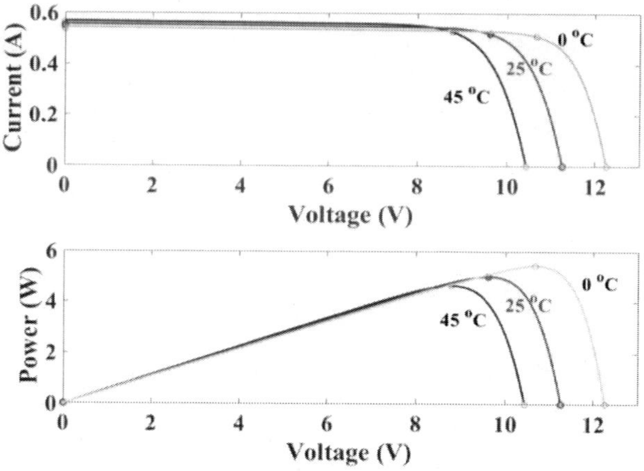

Figure 4. I-V and P-V plot under specified temperatures at $1000 W/m^2$ for a 5W module.

Total Cross-Tied (TCT)

TCT SPVA configuration ranks best among the existing conventional topologies. It is derived from the S-P connection where it is seen that TCT has several solar modules connected in series and parallel orientation. It involves cross-ties. Figure 5 (c) shows 4 × 4 TCT connected SPVA where 4 solar module strings are connected in parallel. Such an arrangement is then linked in a series fashion. The current flowing in individual strings adds up to 4 solar module currents. Thus, the total output current for the considered 4 × 4 TCT connected SPVA is equivalent to four times the module current (I_m). Similarly, the total output voltage for the 4 × 4 TCT connected SPVA is four times the module voltage (V_m). About other conventional topologies, TCT improves ɳ and FF to a greater extent as compared to S-P and B-L but as it consists of many connections, hence this makes the system complex, increasing cost and wiring loss [5]. Figure 6 shows a P-V and I-V plot of 4 × 4 TCT-connected SPVA under various PSC.

The SPVA current (I_{SPVA}), voltage (V_{SPVA}), and power (P_{SPVA}) are represented by equations (9), (10), and (11).

$$I_{SPVA} = 4I_m \quad (9)$$

$$V_{SPVA} = 4V_m \quad (10)$$

$$P_{SPVA} = V_{SPVA}I_{SPVA} = 16V_mI_m \quad (11)$$

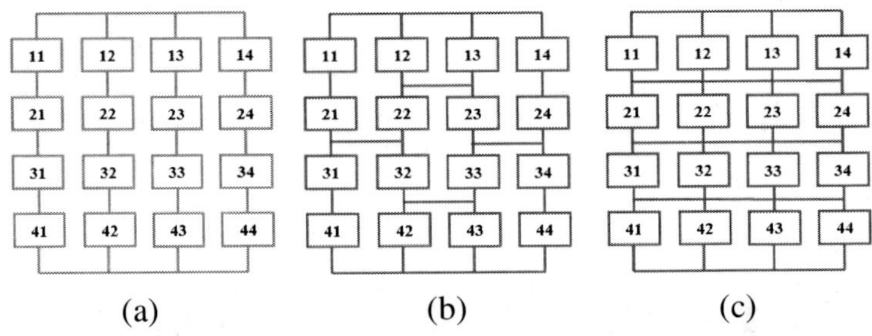

Figure 5. Schematic representation of 4×4 (a) S-P (b) B-L (c) TCT connected SPVA.

Advanced Reconfiguration Topologies

This sub-section discusses a few advanced reconfigured topologies as follows:

Ken-Ken Reconfiguration (K-K)

K-K comes with the logic of SDK comprising of several grids arranged horizontally and vertically adhering to certain rules of the puzzle. Figure 7 (a) shows the fashion in which the interconnections are incorporated for this particular pattern where numbers are assigned from 1 to 4 in respective grids by performing the basic mathematical operations. Each grid is allotted with a particular number and a mathematical operation symbol respectively. There should not be any repetition in the number for any specific horizontal and vertical arrangements. Apart from improvement in GMPP and other parameters, K-K involves many ties compared to TCT-connected SPVA, hence cost, wiring losses, and complexity in structure increase [18].

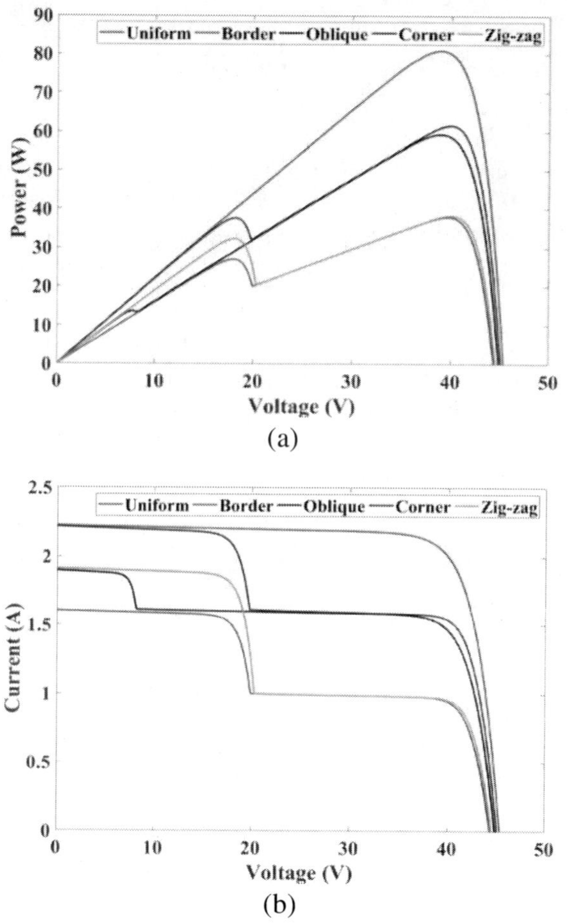

Figure 6. (a) P-V (b) I-V curve under various shading patterns for 4×4 TCT.

Arithmetic Sequence Reconfiguration (AS)

In this advanced reconfigured topology, the physical locations of the SPVA are altered without disturbing the electrical connections. Here, each grid of the SPVA is numbered in a distinct way such that even in extreme cases when shadow condition occurs, it is capable of dispersing the shadow over complete SPVA and each cell receives optimum current during PSC. The layout of AS pattern is shown in Figure 7 (b). It can be observed from Figure 7 (b) that physical connections have been altered in a fashion that each sub-module is specially coded in itself. This topology can be implemented for both symmetrical and asymmetrical SPVA. This advanced reconfigured topology offers several advantages as it increases global power, minimizes power loss, enhances FF, and Ƞ. It has several interconnections leading to a complex circuit design, thus increasing the cost of cables [31].

L-Shape Reconfiguration (L-S)

The factor to be paid attention to for L-S is the initial node (horizontal grid) and the respective final or previous nodes (vertical grid). It proceeds with the formation of the alphabet 'L' from the initial node to the final or previous node. L-S must halt on reaching the final or previous node and continue again with the new vertical node. This procedure continues until all the solar cells get paired at an optimized distance horizontally or vertically. Rearrangement of 4 × 4 L-S connected SPVA is considered in this study as shown in Figure 7 (c). Figure 8 represents I-V and P-V curves under different shading scenarios for L-S connections respectively [28].

41	12	33	24
21	32	43	14
11	42	23	34
31	22	13	44

(a)

31	42	13	24
41	12	23	34
11	22	33	44
21	32	43	14

(b)

11	23	32	44
21	33	42	14
31	43	12	24
41	13	22	34

(c)

Figure 7. Grouping for (a) K-K (b) AS (c) L-S.

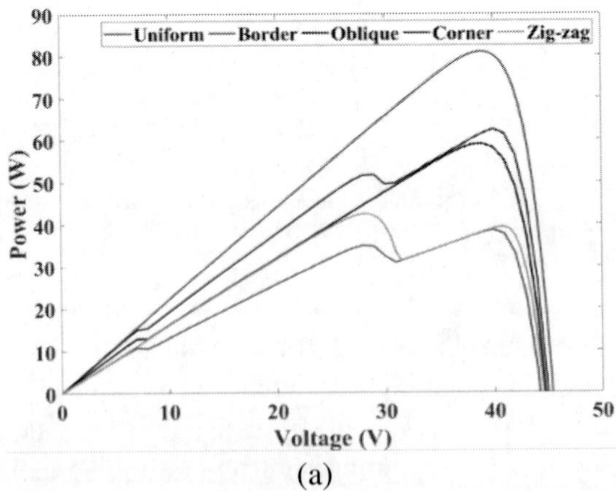

(a)

Figure 8. (Continued).

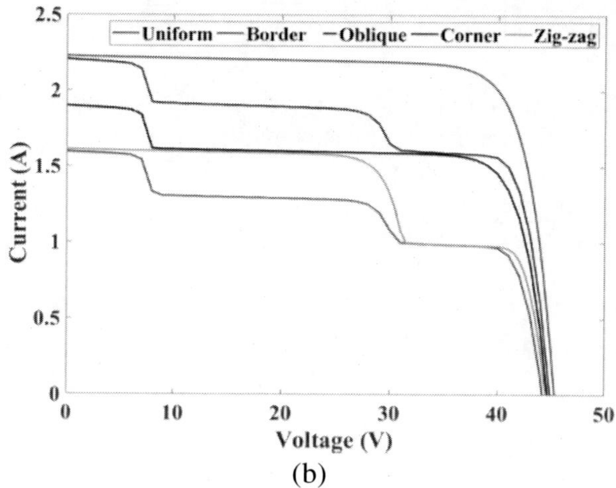

(b)

Figure 8. (a) P-V (b) I-V curve under various shading patterns for 4×4 L-S.

Performance Indices under PSC

Following are a few performance indices that are considered under PSC.

Global Maximum Power Point (GMPP)

It is the maximum point on the P-V plot where power tracked is maximum under PSC.

Efficiency (η)

It is the ratio of maximum power to the product of the area of the SPVA module and the irradiation level falling on the SPVA module. It is formulated by equation (12) [31].

$$\eta = \frac{\text{GMPP}}{\text{Area of SPVA module} \times \text{irradiation incident on SPVA module}} \tag{12}$$

Fill Factor (FF)

It is calculated by equation (13). To yield accurate execution, it is desired that the value of the fill factor is closer to the unity [19].

$$FF = \frac{GMPP(\text{at PSC})}{V_{OC} \times I_{SC}} \tag{13}$$

% Power Loss (%PL)

It denotes the amount of power loss under PSC and is calculated with the help of equation (14) [19].

$$\%PL = \frac{GMPP_{(STC)} - GMPP_{(PSC)}}{GMPP_{(STC)}} \times 100 \tag{14}$$

Mismatch Loss (ML)

It is the difference between maximum power obtained at standard test conditions and partial shading conditions and is evaluated by equation (15) [31].

$$ML = GMPP_{(STC)} - GMPP_{(PSC)} \tag{15}$$

Execution Ratio (ER)

It is defined as the ratio of global peak power at PSC to global peak power at standard test conditions and is calculated by equation (16) [15].

$$ER = \frac{GMPP_{(PSC)}}{GMPP_{(STC)}} \times 100 \tag{16}$$

Result and Discussion

Global Maximum Power Point (GMPP)

For a fully illuminated scenario, GMPP is observed to be 80W both for TCT and L-S. When PSC arises, GMPP under the border, oblique, corner, and zig-zag conditions are 37.89W, 59.24W, 69.51W, and 38.24W for TCT connected SPVA and 38.58W, 58.93W, 62.40W, and 42.40W for L-S at 450W/m^2 and 25°C as shown in Figure 9.

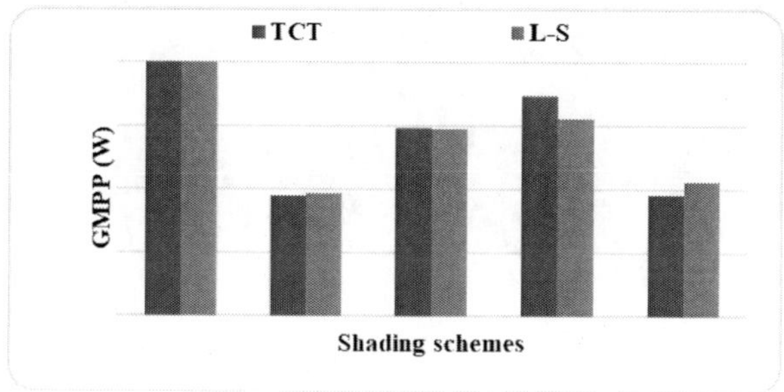

Figure 9. GMPP illustration under PSC.

Efficiency (ꞃ)

For a fully illuminated scenario, ꞃ is observed to be 15.96% both for TCT and L-S. When PSC arises, ꞃ under border, oblique, corner and zig-zag condition are 12.72%, 15.39%, 14.07%, and 11.49% for TCT connected SPVA and 12.95%, 15.31%, 14.27%, and 12.74% for L-S at 450W/m2 and 25oC as shown in Figure 10.

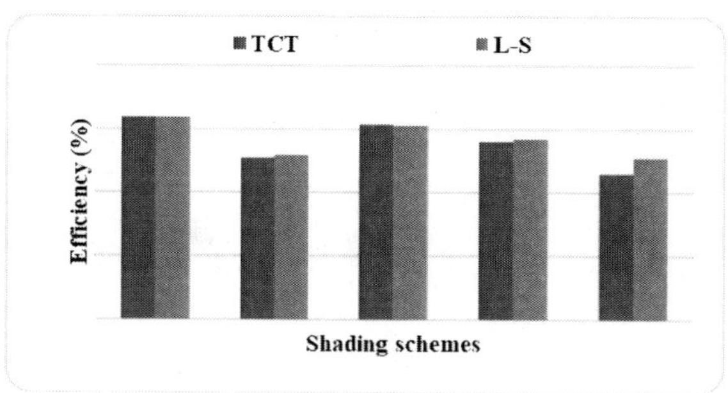

Figure 10. Efficiency illustration under PSC.

Fill Factor (FF)

For a fully illuminated scenario, FF is observed to be 80.05% both for TCT and L-S. When PSC arises, FF under border, oblique, corner, and zig-zag condition are 53.10%, 69.44%, 61.46%, and 44.81% for TCT connected SPVA and 53.76%, 68.94%, 62.90%, and 58.47% for L-S at 450W/m2 and 25oC as shown in Figure 11.

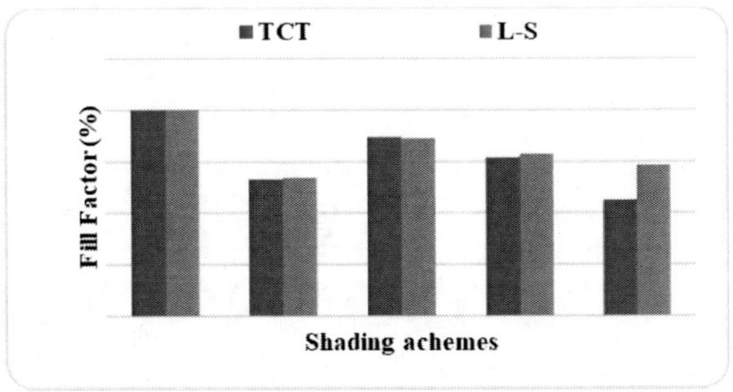

Figure 11. Fill factor illustration under PSC.

% Power Loss (%PL)

For a fully illuminated scenario, %PL is observed to be nil both for TCT and L-S. When PSC arises, %PL under border, oblique, corner, and zig-zag condition are 53.17%, 26.78%, 23.97%, and 52.73% for TCT connected SPVA and 52.31%, 27.15%, 22.87%, and 47.59% for L-S at 450W/m^2 and 25°C as shown in Figure 12.

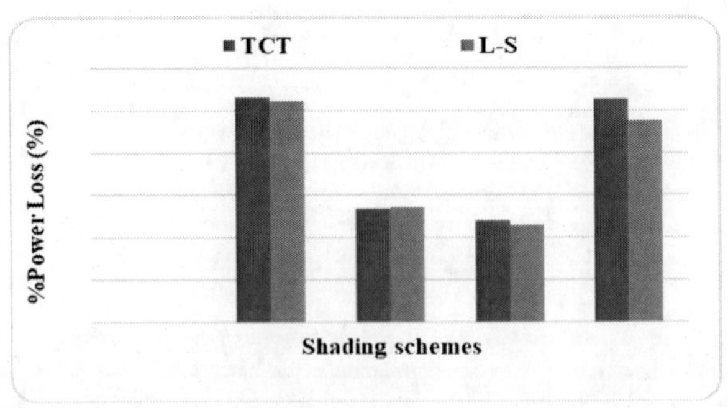

Figure 12. %Power loss illustration under PSC.

Mismatch Loss (ML)

For a fully illuminated scenario, ML is observed to be nil both for TCT and L-S. When PSC arises, ML under the border, oblique, corner, and, zig-zag conditions are 43.01W, 21.66W, 19.38W, and 42.66W for TCT connected SPVA and 42.32W, 21.96W, 18.50W, and 38.50W for L-S at 450W/m^2 and 25°C as shown in Figure 13.

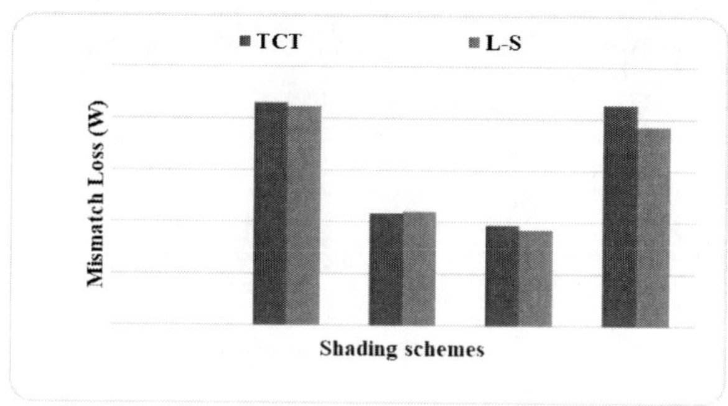

Figure 13. Mismatch loss illustration under PSC.

Execution Ratio (ER)

When PSC arises, under border, oblique, corner, and zig-zag condition are 46.83%, 73.22%, 76.03%, and 47.26% for TCT connected SPVA and 47.68%, 72.85%, 77.13%, and 52.41% for L-S at 450W/m^2 and 25°C as shown in Figure 14.

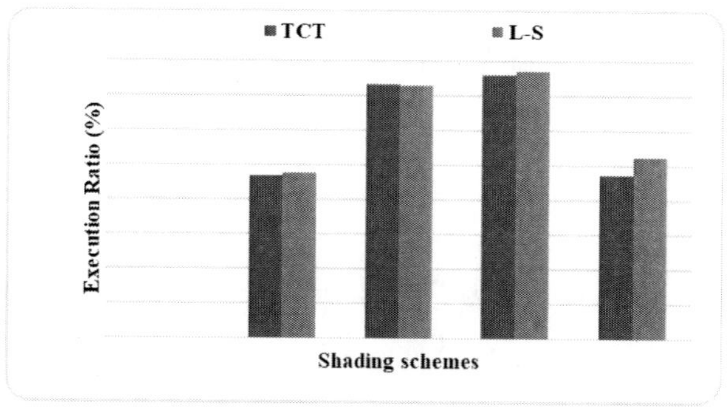

Figure 14. Execution ratio illustration under PSC.

Comparison of TCT and L-S

Table 2 represents a comparison between TCT and L-S. It is observed that L-S under most of the considered shading scenarios is capable of performing effectively and with increased accuracy. The considered performance parameters are enhanced for L-S and it is capable of effective shade-dispersion over the entire SPVA.

Table 2. Comparative analysis of TCT and L-S

Performance parameters	TCT	L-S	Major observations	Conclusion
GMPP	Fig. 9		L-S has maximum power tracked for border and zig-zag shading at 38.58W and 42.40W respectively concerning TCT. TCT is superior to L-S in producing maximum power for oblique and corner shading at 59.24W and 69.51W respectively.	For respective shading scenarios, both TCT and L-S are capable of tracking GMPP. Hence both can be considered good in their respective output for corresponding shadows.
η	Fig. 10		For the border, corner, and zig-zag shading, L-S is superior to TCT as L-S has higher η of 12.95%, 14.27%, and 12.74% respectively. For oblique shading, η is greater than L-S at 15.39%.	L-S is preceding TCT for the major considered shading patterns.
FF	Fig. 11		For the border, corner, and zig-zag shading, L-S is superior to TCT as L-S has higher FF of 53.76%, 62.9%, and 58.47% respectively but for oblique shading TCT is much better at 69.44%.	L-S is much more feasible in shade dispersion than TCT for the major considered shading patterns.
%PL	Fig. 12		Major loss in power is observed for TCT under all shading patterns.	L-S is best as it is capable of minimizing the %PL for all 4 considered shadings.
ML	Fig. 13		TCT is better under oblique shading only as it generates ML of 21.66W whereas for the rest of the considered shading L-S outruns TCT as it generates less ML of 42.32W, 18.50W, 38.50W.	L-S is much more capable and far better than TCT in reducing ML.
ER	Fig. 14		ER is higher for L-S under all the shadings except for oblique which is 73.22% for TCT.	L-S has increased values of ER for a maximum number of shading patterns under consideration than TCT.

Conclusion

Partial shading condition is one of the most intricate issues observed in the current scenario which not only hampers the peak power generation but also leads to mal-function of SPVA and under most severe cases it might lead to fire damaging the entire SPVA installation. Hence to address such an issue, an effort has been made in this chapter to introduce some mitigation steps and one of them is a reconfiguration of SPVA. This chapter focuses on the analysis of conventional models such as S-P, B-L, and TCT where TCT has been much more detailed out and compared with the advanced reconfigured model (L-S). Further, two more advanced reconfiguration has been introduced in this chapter namely K-K and AS. TCT under conventional class and L-S under advanced reconfigured class has been compared for evaluation of GMPP, Ƞ, FF, %PL, ML, and ER under various shading scenarios. The following are some important conclusions for this chapter.

- Maximum power is generated by L-S for border and zig-zag shading at 38.58W and 42.40W respectively for TCT whereas TCT is superior to L-S in producing maximum power for oblique and corner shading at 59.24W and 69.51W respectively.
- L-S is far better than TCT as L-S has higher Ƞ of 12.95%, 14.27%, and 12.74% respectively for the border, corner, and zig-zag shading. For oblique shading Ƞ of TCT exceeds L-S at 15.39%.
- L-S is superior to TCT as L-S has higher FF of 53.76%, 62.9%, and 58.47% respectively for the border, corner, and zig-zag shading, but for oblique shading, TCT is much better at 69.44% than L-S for the same shading pattern.
- L-S lessens ML than TCT and is observed to be more feasible.
- L-S has increased values of ER for a maximum number of shading patterns under consideration than TCT.

Hence overall L-S is observed to be the best in shade dispersion, maximum GMPP generation, and reducing %PL and ML to a substantial amount. This is the superiority of advanced-based reconfiguration models over conventional ones.

References

[1] Pillai, D.S., Ram, J. P., Nihanth, M. S. S. and Rajasekar, N., (2018). A simple, sensor-less and fixed reconfiguration scheme for maximum power enhancement in PV systems. *Energy Convers. Manag*, 402–417.

[2] Mohapatra, A., Nayak, B. and Mohanty, K. B., (2021). Analytical approach to locate multiple power peaks of the photovoltaic array under partial shading conditions and hybrid array configuration schemes to reduce mismatch losses. *Energy Sources, Part A Recover. Util. Environ. Eff.*, 1–22.

[3] Pradhan, R. and Kar, S., (2020). A Comprehensive Study of Partial Shading Effect on the Performance of PV array with Different Configuration. *Proc. 2020 Int. Conf. Renew. Energy Integr. into Smart Grids A Multidiscip. Approach to Technol. Model. Simulation, ICREISG,* 78–83.

[4] Satpathy, P. R. and Sharma, R., Power and mismatch losses mitigation by a fixed electrical reconfiguration technique for partially shaded photovoltaic arrays. *Energy Convers. Manag.*, 52–70.

[5] Bingöl, O. and Özkaya, B., (2017). *Analysis and comparison of different PV array configurations under partial shading conditions*. 336–343.

[6] Ye, C. E., Tai, C. C., Huang, Y. P., and Chen, J. J., (2021). Dispersed partial shading effect and reduced power loss in a PV array using a complementary SuDoKu puzzle topology. *Energy Convers. Manag.*, 114675.

[7] Sagar, G., Pathak, D., Gaur, P, and Jain, V., (2020). A Su Do Ku puzzle-based shade dispersion for maximum power enhancement of partially shaded hybrid bridge-link-total-cross-tied PV array. *Sol. Energy*, 161–180.

[8] Dhanalakshmi, B. andRajasekar, N., (2017). Dominance square-based array reconfiguration scheme for power loss reduction in solar Photo-Voltaic (PV) systems. *Energy Convers. Manag.*, 84–102.

[9] Manjunath, Suresh, H. N. and Rajanna, S., (2019). Performance enhancement of Hybrid interconnected Solar Photovoltaic array using shade dispersion Magic Square Puzzle Pattern technique under partial shading conditions. *Sol. Energy*, 602–617.

[10] Reddy, S. and Yammani, C., (2020). A novel Magic-Square puzzle-based one-time PV reconfiguration technique to mitigate mismatch power loss under various partial shading conditions. *Optik (Stuttg).* 165289.

[11] Varma, G. H. K., Barry, V. R. and Jain, R. K., (2021). A Novel Magic Square Based Physical Reconfiguration for Power Enhancement in Larger Size Photovoltaic Array. *IETE J. Res.* DOI: 10.1080/03772063.2021.1944333.

[12] Venkateswari R. and Rajasekar, N., (2020). Power enhancement of PV system via physical array reconfiguration based on Lo Shu technique. *Energy Convers. Manag.*, 112885.

[13] Nihanth, Malisetty Siva Sai, J. Prasanth Ram, Dhanup S. Pillai, Amer M.Y.M. Ghias, Akhil Garg, and N. Rajasekar. (2019). Enhanced power production in PV arrays using a new skyscraper puzzle-based one-time reconfiguration procedure under partial shade conditions (PSCs). *Sol. Energy*, pp. 209–224.

[14] Krishna, G. S., and Moger, T., (2020). Static Reconfiguration approach for Photovoltaic Array to Improve Maximum power. *Int. Conf. Electr. Electron. Eng.*, 270–275.

[15] Palpandian, M. and David, P. W., (2021). *A Jigsaw Puzzle-based Reconfiguration Technique for Enhancing Maximum Power in Partial Shaded Hybrid Photovoltaic Array.* DOI: 10.21203/rs.3.rs-513975/v1.

[16] Yadav, K., Kumar, B.and Swaroop, D., (2020). Mitigation of Mismatch Power Losses of PV Array under Partial Shading Condition using novel Odd-Even Configuration. *Energy Reports*, 427–437.

[17] Rezazadeh, Sevda, Arash Moradzadeh, Seyed Majid Hashemzadeh, Kazem Pourhossein, Behnam Mohammadi-Ivatloo, and Seyed Hossein Hosseini. (2021). A novel prime numbers-based PV array reconfiguration solution to produce maximum energy under partial shade conditions. *Sustain. Energy Technol. Assessments.* 47(1):101498.

[18] Palpandian, Murugesan, David Prince Winston, Balachandran Praveen Kumar, Cherukuri Santhan Kumar, Thanikanti Sudhakar Babu, and Hassan Haes Alhelou, (2021). A New Ken-Ken Puzzle Pattern Based Reconfiguration Technique for Maximum Power Extraction in Partial Shaded Solar PV Array. *IEEE Access*, 65824–65837.

[19] Kumar Pachauri, R., Thanikanti, S. B., Bai, J., Kumar Yadav, V., Aljafari, B., Ghosh, S., & Haes Alhelou, H. (2022). Ancient Chinese magic square-based PV array reconfiguration methodology to reduce power loss under partial shading conditions. *Energy Convers. Manag.* 253, 115148.

[20] Adav, A. S., Pachauri, R. K., Chauhan, Y. K., Choudhury, S., & Singh, R. (2017).Performance enhancement of partially shaded PV array using novel shade dispersion effect on magic-square puzzle configuration. *Sol. Energy*, 780–797.

[21] Dhanalakshmi B. and Rajasekar, N. (2018). A novel Competence Square-based PV array reconfiguration technique for solar PV maximum power extraction. *Energy Convers. Manag.*, 897–912.

[22] Madhusudanan, Gurusamy, Subramaniam Senthilkumar, I. Anand, and Padmanaban Sanjeevikumar. (2018). A shade dispersion scheme using Latin square arrangement to enhance power production in a solar photovoltaic array under partial shading conditions, *J. Renew. Sustain. Energy.*, 10(5).

[23] Mansur, A. Al and Amin, M. R., (2019). Performance investigation of different PV array configurations at partial shading conditions for maximum power output. *Int. Conf. Sustain. Technol. Ind.,* 1–5.

[24] Krishna, G. S., and Moger, T. (2019). Improved SuDoKu reconfiguration technique for total-cross-tied PV array to enhance maximum power under partial shading conditions. *Renew. Sustain. Energy Rev.*, 333–348.

[25] Satpathy, P. R. and Sharma, R., (2019). Diffusion charge compensation strategy for power balancing in capacitor-less photovoltaic modules during partial shading. *Appl. Energy*, 113826.

[26] Alkahtani, Mohammed, Zuyu Wu, Colin Sokol Kuka, Muflah S. Alahammad, and Kai Ni. (2020). A Novel PV Array Reconfiguration Algorithm Approach to

Optimising Power Generation across Non-Uniformly Aged PV Arrays by Merely Repositioning. *J — Multidiscip. Sci. J.*, 32–53.

[27] Sharma, Snigdha, Lokesh Varshney, Rajvikram Madurai Elavarasan, Akanksha Singh S. Vardhan, Aanchal Singh S. Vardhan, R. K. Saket, Umashankar Subramaniam, and Eklas Hossain (2021). Performance Enhancement of PV System Configurations under Partial Shading Conditions Using MS Method. *IEEE Access*, 56630–56644.

[28] Rasheed, M., Alabdali O., and Shihab S., (2021). A New Technique for Solar Cell Parameters Estimation of the Single-Diode Model. *J. Phys. Conf. Ser.*, 1879(3):032120.

[29] Bayoumi, Ahmed S., Ragab A. El-Sehiemy, Karar Mahmoud, Matti Lehtonen, and Mohamed M. F. Darwish. (2021). Assessment of an Improved Three-Diode against Modified Two-Diode Patterns of MCS Solar Cells Associated with Soft Parameter Estimation Paradigms. *Applied Sciences* 11(3):1055.

[30] Singh, K., Chandel, T. A., Mallick, M. A. and Siddiqui, M. K. (2019). Effect of bypass diode under partial shading in SPV module. *Int. J. Eng. Adv. Technol.*, 2215–2219.

[31] Anjum, S., and Mukherjee, V., (2021). A novel arithmetic sequence pattern reconfiguration technique for line loss reduction of the photovoltaic array under non-uniform irradiance. *J. Clean. Prod.*, 129822-129849.

Chapter 8

Implementation of Metaheuristic MPPT Approaches for a Large-Scale Wind Turbine System

Diwaker Pathak[1,*], Aanchal Katyal[2,†], Prerna Gaur[2,‡] and Yogesh Kumar Chauhan[3,§]

[1]Department of Electrical Engineering, Teerthankar Mahaveer University, Moradabad, India

[2]Instrumentation and Control Engineering Department, Netaji Subhas University of Technology, New Delhi, India

[3]Department of Electrical Engineering, Kamla Nehru Institute of Technology, Sultanpur, Uttar Pradesh, India

Abstract

The wind turbine system (WTS) has a major challenge of highly variable wind speed. The maximum available power corresponding to each value of wind speed needs to be extracted at each variation for the efficient operation of the WTS. This assessment is achieved using a maximum power point tracking (MPPT) method. In this research work, the maximum available power in a 2-MW WTS is effectively tracked at each instant of wind variations using various metaheuristic algorithms so that the capital cost of the WTS can be achieved in a minimum period. Firstly, a 2-MW wind turbine model is modeled in MATLAB/Simulink and characteristics of power coefficient versus tip speed ratio (TSR) and

---

[*]Corresponding Author's Email: diwakergnit29@gmail.com.
[†]Corresponding Author's Email: aanchalk.ie20@nsut.ac.in.
[‡]Corresponding Author's Email: prernagaur@yahoo.com.
[§]Corresponding Author's Email: chauhanyk@yahoo.com.

In: Applied Artificial Intelligence (AI) to Green Power Technology
Editors: Yogesh Kumar Chauhan, Ranjan Kumar Behera and Asheesh K. Singh
ISBN: 979-8-88697-131-6
© 2022 Nova Science Publishers, Inc.

turbine power versus turbine rotor speed are achieved to decide the optimal points for the effective MPPT. Secondly, in strategy to maximize the turbine power at the corresponding wind speeds, metaheuristic algorithms such as whale optimization (WOA), grasshopper optimization (GOA), gray wolf optimization (GWO), differential squirrel search algorithm (DSSA), and hybrid particle swarm optimization (PSO)-GWO are implemented. Wind speeds are taken as input to each MPPT algorithm and corresponding optimal rotor speeds of the turbine are determined so that corresponding maximum power can be extracted at a pre-defined optimal TSR value. Furthermore, all the metaheuristic algorithms are well tested at various benchmark functions and exploration, exploitation, and convergence analysis have been carried out for the performance comparison. The MPPT performance is assessed on the power coefficient versus TSR curve that shows the effective tracking of MPP point by point in a simple way.

Keywords: MPPT control, wind turbine, AI, metaheuristic algorithms, GWO, DSSA

Introduction

In various engineering applications, renewable energy sources (RESs) such as wind and solar energy infiltrated strappingly as the use of conventional energy sources like coal, petroleum, etc. has many environmental apprehensions (Pathak and Gaur 2019). Among all the RESs, wind energy has grown most rapidly (Sagar et al. 2020). In a wind turbine system (WTS), the kinetic energy of the wind is directly converted into mechanical energy using the wind turbine (WT) coupled to the electrical generator by means of a shaft and gear-box. This generated mechanical energy acts like a prime mover and is used to rotate the shaft of the generator to produce electrical energy while the load is connected to the stator of the generator. The generation and control of turbine output mechanical power can be divided into three specific ranges of wind speed: power generation starts from the cut-in wind speed ($V_{wcut\text{-}in}$) and increases with the increase in wind speed up to the rated value (P_{rated} at $V_{w\text{-}rated}$); once the rated mechanical power is reached, power should not go beyond the rating of the WTS, hence, WT power should be maintained constant up to the cut-out speed ($V_{wcut\text{-}out}$). Beyond the $V_{wcut\text{-}out}$, the prime mover should immediately be taken to the stationary position to meet the safe operation of WTS.

Based on the different wind generator topologies, the WTSs are characterized as variable speed WT (VSWT) and fixed speed WT (FSWT). The FSWT has various disadvantages such as having the heavy mechanical strain, small operating range due to fixed speed (±5%), and consequently, a heavy gear-box with multiple stages is essential to meet the load demand that causes the large fluctuations in output power (Pathak and Gaur 2020). VSWT has various advantages over the FSWT (Mousa, Youssef, and Mohamed 2019,) (Wang, He, Wang, Zhang, 2022). Out of them, the foremost ease by the VSWT is that, for each value of wind speed, maximum turbine power can be extracted with a reduced mechanical strain and output power fluctuations. The maximum power extraction can be achieved using various techniques espoused by different maximum power point tracking (MPPT) controllers those majorly can be characterized into four types: tip speed ratio (TSR) MPPT control, power signal feedback (PSF) MPPT control, optimal torque control (OTC) MPPT method and hill-climbing search (HCS) MPPT methods (Zouheyr, Lotfi, Abdelmadjid, 2021).

The OTC MPPT method comes under the indirect power control (IPC) and encompasses the torque regulation of the wind generator according to the reference torque corresponding to the maximum power for various wind speeds (Mousa, Youssef, and Mohamed 2019). The intrinsic drawback is that wind speed is not measured directly, hence, the wind speed variation may remain disregarded in the reference signal. Other disadvantage is of being dependent on the environmental conditions and the necessity of overall knowledge about WT characteristics.

The next MPPT control under the IPC is power signal feedback (PSF) control. In this strategy, the foremost requirement is a lookup table of the various optimal power curves by rigorous simulations for the particular WT (Mousa, Youssef, and Mohamed 2019). The rotor speed value is calculated by looking at the optimal power curves and desired electrical power is achieved. Although the PSF and OTC MPPT methods are low-cost and robust, the downsides are nearly analogous for WT with large inertia at the lower values of the wind speeds. The most commonly implemented MPPT for the VSWT is TSR MPPT control, where each wind speed value and corresponding rotor speed value is used to estimate the reference value of rotor speed. All the measured rotor speeds corresponding to each wind speed are then compared to the reference rotor speed to maintain the constant power coefficient at the optimal value of TSR. Being simple structured and parameter sensitive, this MPPT method offers the highest turbine efficiency, less oscillation in output power and faster convergence speed as compared to other IPC schemes as TSR is a constant value. The only issue with this method is the high initial implementation cost.

The direct power control (DPC) schemes to track the MPP includes perturb and observe (P&O), incremental conductance (INC) and optimal relation-based (ORB) MPPT methods. The conventional P&O algorithm has a major advantage of excluding the sensor requirement like an anemometer (Mousa, Youssef, and Mohamed 2019). On the other hand, the problem with this algorithm is facing the challenge of deciding a suitable step size for perturbation that becomes a major parameter affecting the WTS performance. In case of larger step size, oscillation around the zero slopes will be large and for smaller step size, convergence time and settling response will have deteriorated. INC MPPT method needs reference values of DC-link parameters to calculate the optimal power and tracks the MPPT accordingly. ORB MPPT method again requires various information from the WTS like rotor speed, mechanical torque, DC-link voltage, and current to calculate the optimal power that is later compared to the actual value of mechanical power and error signal is given to the controller to generate the required pulses for the power converter. Therefore, the major challenge in the DPC MPPT methods is the larger oscillations near the MPPT that in turn reduce the efficiency of the WTS.

Such complications are perceived by researchers in recent years, with the design of artificial intelligence (AI) and mathematical optimization methods along with the modified and hybrid MPPT frameworks to deal with this problem. The authors in (Kumar and Chatterjee 2016), have carried out a detailed survey on the various MPPT methods and concluded that IPC MPPT methods are most effective, promising, and reliable if analyzed using the adaptive and AI-based metaheuristic algorithms. Another AI-based MPPT method was proposed in (Lee and Kim 2016) using the fuzzy logic (FL) based P&O algorithm. Using the FL approach, the step size of the convention P&O MPPT was varied and overall performance was enhanced as compared to the basic one. In (Sitharthan et al. 2020), the authors have developed a new PSO-based RBFNN supported TSR MPPT approach that improves the reliability of WTS and offers an efficient power tracking capability. (Priyadarshi et al. 2019) analyses the application of the ant colony optimization (ACO) algorithm for the effective MPPT control. It was concluded that the ACO MPPT method has seven times the faster-tracking speed as compared to the PSO MPPT. These metaheuristic algorithms based MPPT methods were also examined in real-time platform to show the simple implementation. The authors in (Mokhtari and Rekioua 2018), the ACO algorithm is used to regulate the PI controller parameters that govern the MPPT. It was concluded that using the ACO, the power coefficient can be maintained at its maximum value, therefore, extracting the maximum power continuously. The authors in (Fathy and El-Baksawi 2019), developed a new MPPT method based on the grasshopper optimization algorithm (GOA) for VSWT. The performance of GOA MPPT was compared to the PSO, cuckoo search (CS) and crow search algorithm (CSA) based MPPT methods and found that GOA provides better convergence speed and minimum oscillations around the MPP. The authors in (Ali and Ouassaid 2019), have examined the gray wolf optimization (GWO) based TSR MPPT technique and the obtained performance is compared with the PSO based TSR MPPT method for a large scale simple model of WT. It was found that GWO based TSR MPPT outperforms the PSO-based TSR MPPT method in terms of tracking speed, power oscillation, settling time, etc. However, the exploration, exploitation, and convergence analysis are needed to be performed with respect to the various benchmarks for the analyzed MPPT method that ensures effective performance.

Although the numerous described approaches are available to simulate the MPPT with WTS, the metaheuristic algorithm-based applications are still inadequate and require more consideration particularly after they are proven to have higher efficiency to achieve the MPP of other nonlinear systems like photovoltaic systems (Rezk, Fathy, and Abdelaziz 2017) and fuel cell (Fathy et al. 2021). The implementation and the overall operation of the metaheuristic MPPT approach is needed to be better explained based on the optimal characteristics of the WT.

To cover up the inconsistency that is presently owing to the preceding MPPT approaches, an effective and powerful metaheuristic approach is required to track the MPPT. Further, this required MPPT approach is needed to be presented in a simple way that explains the overall circumstance of tracking the global maxima of the WT characteristic. Therefore, in this chapter, a simple model of a large-scale WTS is considered and some new applications of the metaheuristic algorithms based MPPT are examined and explained in a basic manner. A new application of the hybrid GWO-PSO MPPT algorithm is proposed along with the differential squirrel search algorithm (DSSA) and obtained performance is compared with the GWO, GOA, and WOA algorithms based TSR MPPT methods. On the mathematical model of WTS, every optimization method is well-tested on standard benchmark unimodular and multi-modular functions, and exploration, exploitation, and convergence analysis are assessed to show the effectiveness. Furthermore, the balance between the exploitation and exploration segments is established that aids the method to attain the global maxima effectively. The contribution of the current work is summarized in the following points:

1. The application of the hybrid GWO-PSO algorithms is implemented for the TSR MPPT control of a WTS as a first-ever approach.
2. DSSA algorithm is applied to control the TSR of a WTS for the MPPT for the first time.
3. Exploration and exploitation analysis has been carried out for the DSSA and hybrid GWO-PSO algorithms before applying for the TSR MPPT process.
4. A simple approach for the TSR MPPT method is presented to observe the global maxima search.

System Description and Modeling

Wind Turbine Model

The WT model is presented in Figure 1. Using the well-known equations described in this subsection, turbine/mechanical torque is calculated and the required rotor speed is generated (Pathak, Bhati, and Gaur 2020). To convert the kinetic energy of the wind to mechanical energy, WTs are used. This mechanical power is given by the following expression:

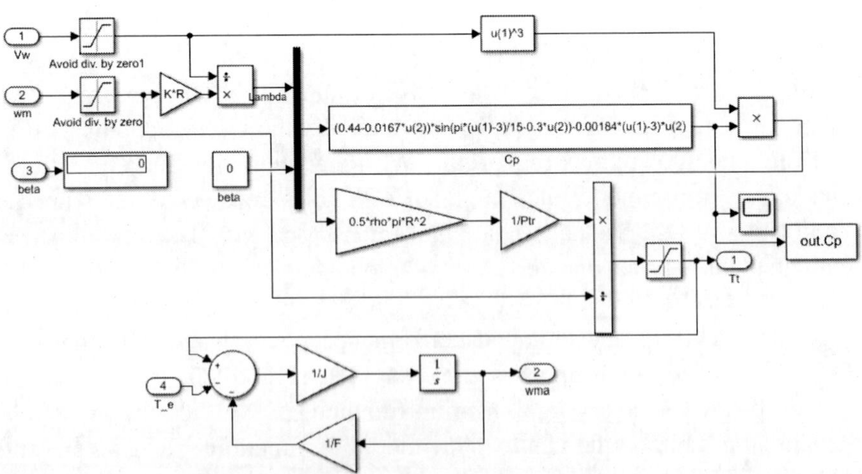

Figure 1. Mathematical representation of wind turbine system.

$$P_m = C_p(\lambda, \beta) \cdot \left[\frac{1}{2} \rho \pi R^2 V_w^{\ 3} \right] \tag{1}$$

where ρ is air density (1.225 kg/m^3), R is WT blade's rotor radius (41 m) and C_p is power coefficient. C_p mainly depends upon pitch angle (β) and tip speed ratio (TSR, λ) and is the ratio of power extracted by a wind turbine to the total power available in the wind.

The pitch angle is kept at 0^0 when power regulation is not required which indicates WT is operating below the rated wind speed and $C_p(\lambda, \beta) \rightarrow C_p(\lambda, 0^0) \rightarrow C_p(\lambda)$. Once wind speed reaches the rated value, power should be regulated at its rated value as per the manufacturer's design parameters (Ali and Ouassaid 2019). To regulate the power at its rating, the pitching angle is increased to shed the energy harvested by the WT or decreased to maximize the power harvesting from the wind. However, with the increase in pitch angle, C_p decreases that indicating WT is running at lower efficiency as shown in Figure 2. TSR is the ratio of the speed of the blade's tips ($\omega_{wt} \times R$) to the wind speed (V_w) and given by (2).

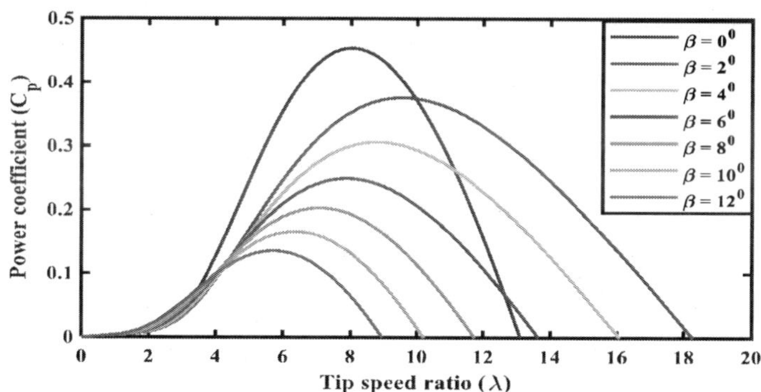

Figure 2. C_p versus λ curve for the understudy system.

$$\lambda = \frac{\omega_{wt} R}{V_w} \tag{2}$$

Hence, any variation in the wind speed affects the rotor speed bringing a variation in the TSR that varies the power coefficient. Figure 2 depicts that there is an optimum WT rotor speed associated with a wind speed. Therefore, there will be an optimum value of TSR at which the WT operates at its maximum efficiency (C_p^{max}), and (2) can be re-written as (3).

$$\lambda_{opt} = \frac{\omega_{wt\_opt} R}{V_w} \tag{3}$$

If the value of λ_{opt} is regulated in such a way that C_p^{max} can be achieved continuously and WT can be operated at its maximum efficiency. This efficiency was proposed by Albert Betz in 1919, known as Betz's coefficient or power coefficient given in (4).

$$C_p(\lambda, \beta) = 0.5176\left(\frac{116}{\lambda_i} - 0.4\beta - 5\right)e^{\left(-\frac{21}{\lambda_i}\right)} - 0.0068\lambda \tag{4}$$

$$\frac{1}{\lambda_i} = \frac{1}{\lambda + 0.08\beta} - \frac{0.035}{\beta^3 + 1} \tag{5}$$

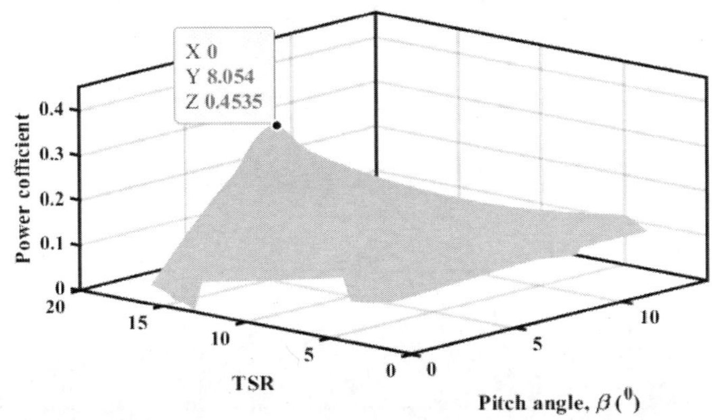

Figure 3. The plot of $C_p(\lambda, \beta)$.

To achieve C_p^{max}, $\beta = 0^0$ should be considered in (4) and (5). Also, from $\frac{dC_p}{d\lambda}\Big|_{\beta=0} = 0$, maximum power coefficient can be calculated along with the point of maxima (λ_{opt}). In this work, λ_{opt} is calculated as 8.054 and C_p^{max} is calculated as 0.4535. The rated wind speed is calculated as 11.2 m/s for the current 2 MW WT. Finally, the turbine/mechanical torque can be calculated from (6) which also depends upon the λ and β. A 3D plot is presented in Figure 3 that shows the variation of C_p in the space offered by λ and β.

$$T_{wt} = \frac{P_m}{\omega_{wt}} = 0.5\rho\frac{C_p(\lambda,\beta)}{\lambda}\pi R^3 V_w^{\ 2} \tag{6}$$

The considered dynamic equation of the rotor in the current work is expressed in (7).

$$J_i\frac{d\omega_{wt}}{dt} = T_e - T_{wt} - F_{vc}\omega_{wt} \tag{7}$$

where F_{vc} is viscous friction coefficient, J_i is the moment of inertia of the WTS, and T_e is electromagnetic torque established by the WG and T_{wt} is the mechanical torque of WT.

Maximum Power Point Tracking

To understand the implemented MPPT operation in the modeled WTS, the operating regions of the WT need to be taken into attention. As discussed earlier, there are three operating regions of a WT as shown in Figure 4. Region 1 is from 0 m/s to 3 m/s of wind speed, where power generation does not take place as the generator requirement is not fulfilled. When wind speed goes beyond the cut-in wind speed (Vw_cut-in, 3 m/s), power increases in cubic relation to the wind speed based on expression (1). A moment comes when wind speed reaches a particular value, where WT attains its power generation at the rated value. This value of wind speed is rated wind speed (Vw_rated). This region from Vw_cut-in to Vw_rated is known as the MPPT region (region 2), where power varies unpredictably according to the wind speed variation. Beyond the Vw_rated in region 3, output power needs to be maintained at the constant value using pitch angle control so that system failure can be prevented and reliability can be ensured. After a critical value of wind speed (Vw_cut-out), WT operation should be stopped (Kurian, Jayanthi and Devraj, 2022).

Being a simple and efficient algorithm for MPPT, the TSR MPPT method is analyzed in this work. This MPPT method calculates a particular rotor speed of the WT for every value of the wind speed at which the maximum power can be calculated and effectively tracked. In this way, the λ_{opt} is maintained constant by varying the rotor speed with variations in the wind speed to ensure the WT operation at MPP as shown in Figure 5. Under the highly varied wind speed profile, the optimum value of rotor speed is determined using (8) which is based on (3).

$$\omega_{wt\_opt} = \left(\frac{\lambda_{opt}}{R}\right) V_w \approx k_{opt} V_w \tag{8}$$

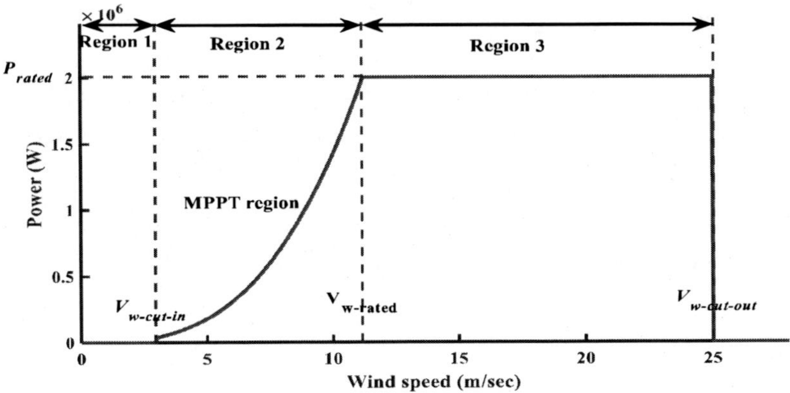

Figure 4. Operating regions for the modeled WTS.

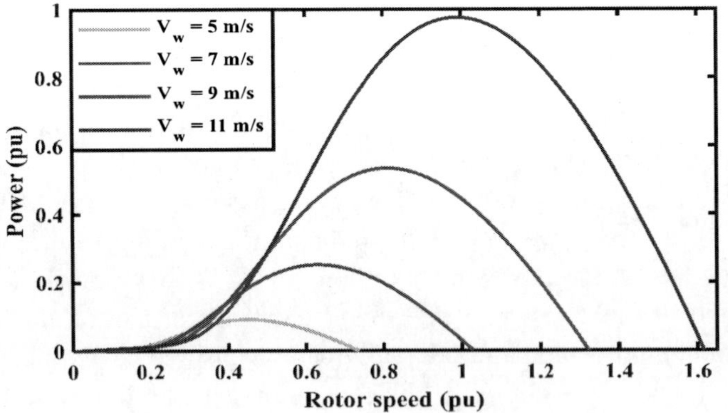

Figure 5. Turbine power-speed curve for the modeled WTS.

With V_w as the wind speed measured by an anemometer, optimal TSR λ_{opt} is calculated as 8.054, and the radius of the swept area by turbine blades R is taken as 41 m. Secondly, optimum rotor speed and actual rotor speeds are regulated using various metaheuristic algorithms. Pseudocode for the TSR MPPT control is given as:

Pseudocode of the TSR MPPT algorithm

```
1. Input wind speed V_w
2. Initialize the rotor speed ω_wt
3. Initialize rotor radius R do
4. Calculate the optimal TSR λ_opt = (ω_wt_opt × R)/V_w or ω_wt_opt = (λ_opt × V_w)/R
5. Calculate power coefficient C_p at β = 0^0
6. for V_w_prev ≤ t ≤ V_w, do
        Regulate the ω_wt to maintain λ_opt at the constant value
    for V_w_prev = V_w, do
        λ_opt_prev = λ_opt
    end for
  end for
  Return Lambda_opt
```

Finally, based on (9), MPP can be achieved by adjusting the ω_{wt} for the WTS. The MPPT algorithm determines the ω_{wt} continuously that maximizes λ to its optimum level, therefore, generating the maximum power.

$$P_{m\_max} = \left[\frac{1}{2} \rho \pi R^2 \left(\frac{\omega_{wt\_opt} R}{\lambda_{opt}} \right)^3 \right] . C_p(\lambda, \beta) \tag{9}$$

WTS Maximum Power Point Tracking Algorithms

In this section, the implementation of some metaheuristic based MPPT algorithms is discussed. The implemented MPPT algorithms have been executed in an infinite loop within the maximum number of iterations provided. When the execution of the MPPT algorithm is completed for all the iterations, the local best value of MPP can be achieved. After the completion of each iteration, the MPPT algorithm assigns the prior value of ω_{wt_opt} at the step where it was calculated as the local best. For the next iteration, the algorithm calls the local best of the previous algorithm and further iterates to find out the global best. Once the ω_{wt_opt} is decided globally, the corresponding point is fixed as a reference value to the WTS model till the wind speed remains constant.

Grey Wolf Optimization Based MPPT Algorithm

GWO simulates major activities like hunting, searching for prey, encircling, and attacking. Each group member is tasked and arranged according to the social hierarchy. The task of hunting is led by a head of wolves ranked as per mental strengths including awareness and supervision, hence, three wolves that have a better perception of the probable location of the victim are labeled as α, β, and δ will guide the whole pack of wolves throughout hunting missions (Mirjalili, Mirjalili, and Lewis 2014). Encircling the prey during hunting can be mathematically expressed in (10) and (11), where i is i^{th} iteration, A and C are coefficient vectors, X is the position vector of a wolf, X_{pos} is the position vector of prey:

$$\vec{E} = \left| \vec{C} . \vec{X}_{pos}(i) - \vec{X}(i) \right| \tag{10}$$

$$\vec{X}(i+1) = \vec{X}_{pos}(i) - \vec{A} . \vec{E} \tag{11}$$

where $\vec{A} = 2\vec{a} . \mu_1 - \vec{a}$ *and* $\vec{C} = 2.\mu_2$. a decreases linearly from 2 to 0 throughout iterations and μ_1 and μ_2 are random vectors in [0, 1]. The position of each wolf is led by α, β, and δ particles and updated using a single equation that combines the hunting, encircling, and attacking as presented in (12), where X_α is the highest priority wolf, X_β and X_δ are lower priority wolf [39].

$$\vec{X}(i+1) = \frac{\left(\vec{X}_\alpha - \vec{A}_1 . \left| \vec{C}_1 . \vec{X}_\alpha - \vec{X} \right| + \vec{X}_\beta - \vec{A}_2 . \left| \vec{C}_2 . \vec{X}_\beta - \vec{X} \right| + \vec{X}_\delta - \vec{A}_3 . \left| \vec{C}_3 . \vec{X}_\delta - \vec{X} \right| \right)}{3} \tag{12}$$

Here, in the case of MPPT, the population of the gray wolf is considered as wind speed. For each wind speed value, the algorithm computes the optimal value of TSR by calculating the rotor speed from the WTS. When the position of X reaches the MPP, the corresponding coefficient vector has a value near zero and gives the highest WT power value (MPP) corresponding to the best value of the TSR in terms of λ_{opt_α}. For an easier understanding, the MPPT implementation is represented in pseudocode as follows:

Pseudocode of GWO algorithm for MPPT

```
Input maximum iterations N.
Initialize position of the pack of the wolf (ω_wt).
Initialize system parameters a, A, and C.
Calculate the fitness of ω_wt.
while (j < N) do
for each wolf do
Update current ω_wt position using (12).
    end for
Update system parameters using (11).
Calculate the fitness of all wolves.
Update global best position of ω_wt to find out maximum power.
    j = j + 1
end while
return X_α
```

Hybrid Particle Swarm Optimization with Grey Wolf Optimization Based MPPT

The best properties of the GWO and particle swarm optimization (PSO) (Lee and Kim 2016) is hybridized to eliminate the shortcomings of the individual PSO and GWO algorithm during the MPPT operation. The hybrid PSO-GWO based MPPT algorithm starts with the initialization of iteration number, upper and lower bounds of $\omega_{wt,}$ and calculates the three best solutions of turbine power using (12) using GWO (Shaheen, Hasanien, and Alkuhayli 2021). Further, as the position of each particle, ω_{wt}, (X_α, X_β and X_δ) is dominated by the velocity of the calculated particle using the PSO commands, the updated position can be calculated as the sum of initial position from the GWO commands and velocity using the PSO algorithm from (13). Therefore, the resultant algorithm executes in a quick manner as the new position of each wolf (particle, ω_{wt}) is now the combination of the best position and velocity of the particle. The global best position of the particle is updated using (14). This hybrid algorithm may be proven a promising one due to its simplicity and effectiveness.

$$velocity(i,j) = w * \begin{pmatrix} velocity(i,j) + c_1 r_1 (X_\alpha - positions(i,j)) + \\ c_2 r_2 (X_\beta - positions(i,j)) + c_3 r_3 (X_\delta - positions(i,j)) \end{pmatrix} \tag{13}$$

Here, c_1, c_2, and c_3 are kept constant. w is the inertia of each wolf.

$$positions\_upd(i,j) = positions(i,j) + velocity(i,j) \tag{14}$$

The hybrid GWO-PSO MPPT algorithm is implemented as an effective method to track the MPP with the effective and quick search of the global solution. The pseudocode of the hybrid PSO_GWO algorithm is presented as follows:

Pseudocode of PSO-GWO algorithm for MPPT

Input maximum iterations N.
Initialize position of the pack of wolfs (ω_{wt}).
Initialize parameters a, A, C, c_1, c_2 and w.
Calculate the fitness of each value of ω_{wt}.
while (j < N) **do**
for each wolf **do**
Update the current position of ω_{wt} in iteration using Eq. (14).
end for
Update system parameters using $\vec{A} = 2\vec{a}.\mu_1 - \vec{a}$ and $\vec{C} = 2.\mu_2$.
Calculate the fitness of ω_{wt} to obtain the maximum power.
Update global best position of ω_{wt}.
j = j + 1
end while
return X_α

Whale Search Optimization Algorithm Based MPPT

Humpback whales execute their hunt for a group of small fishes near the water surface. This hunting is executed by releasing special bubbles gurgled around the boundary. It is found that there are two exercises linked with bubble formation: 'upward-spirals' in which whales dive near about 12 m downward and release bubbles in a spiral manner around the target while reaching to the surface; and the second one is 'double-loops' that executes in three parts (coral loop, lobtail, and capture loop) (Mirjalili and Lewis 2016). In this manner, in the three hunting phases whales executes their hunt: encircling, spiral-shaped bubble release, and then searching for the target.

While encircling the target, whales identify the best search agent that is near the optimal distance. Therefore, the WOA algorithm updates the position of the other search agents according to the position of the best search agent that can be expressed as in (10) and (11) same as in the GWO section. Further, for the second phase, an update on the position and shrinking of encirclement is modeled using (15).

$$\vec{X}_{whale\_pos}(i+1) = \begin{cases} \vec{X}_{prey\_pos}(i) - \vec{A}.\vec{E} & if\ p < 0.5 \\ \vec{D}e^{bl}.cos(2\pi l) + \vec{X}_{prey\_pos}(i) & if\ p \geq 0.5 \end{cases} \tag{15}$$

Pseudocode of WOA algorithm for MPPT

```
Input Maximum iterations N.
Initialize position of the pod of whales (ωwt).
Initialize a, A, C, l and p.
Calculate the fitness of ωwt.
while (j < N) do
for each search agent do
Update current position of ωwt depending on value of probability constant p
using Eq. (15).
end for
Update system parameters using A = 2a.μ1 − a and C = 2.μ2 .
Calculate the fitness for all values of ωwt.
Update global best position of ωwt to calculate the maximum power.
    j = j + 1
end while
return Xα
```

where $\vec{D} = \left| \vec{X}_{prey\_pos}(i) - \vec{X}_{whale\_pos}(i) \right|$ b is constant to decide the algorithmic spiral shape, l and p are randomly chosen between [-1, 1] and [0, 1], respectively. Again, for the third and last phase, the search for the prey can be assessed by the (16) defined in the GWO section.

$$\vec{X}_{whale\_pos}(i+1) = \vec{X}_{rand} - \vec{A}.\vec{E} \tag{16}$$

Differential Squirrel Search Algorithm Based MPPT

DSSA is a hybrid metaheuristic algorithm (Jena et al. 2021) that combines interesting properties of the squirrel search algorithm (SSA) (Jain, Singh, and Rani 2019) and differential evolution algorithm (DE) (Mallipeddi et al. 2011). Based on the SSA model, the search strategy is enhanced by updating the position of the group of squirrels according to the position of the squirrels at a particular tree. Further, the exploration and exploitation properties are enhanced by incorporating the crossover function from the DE. The fitness value of a squirrel in a particular position can be calculated after putting the position value in a fitness function (FF) and then arranged to attain the current best position of the squirrel (P_{sqb}) at the tree. The further three optimal values will be the positions of squirrels at the tree with food P_{sqa} (1:3) that reach towards the best/optimal position in a succeeding iteration.

When the squirrels change their position in the presence of any predatory animal and follow the other squirrels to the safe position. This can be expressed by (17), where P_{sqct}^{old} is the current best of the squirrel, d_g is a random gliding distance, G_c is gliding constant, P_{avg} is the average best position of the squirrel, $\underline{P_{pd}}$ is the probability of predator presence, r_1 is a random number between [0, 1].

$$P_{sqa}^{new} = \begin{cases} P_{sqa}^{old} + d_g G_c \times (P_{sqct}^{old} - P_{sqa}^{old} - P_{avg}), & r_1 \geq P_{pd} \\ Random\ position, & otherwise \end{cases} \tag{17}$$

Now the chances to get trapped at local minima can be avoided by incorporating the crossover function rate C_r at the P_{sqct} from the DE algorithm and the updated new positions can be obtained using (18), where, $j_{rand} \in [1, D]$, $i = 1, 2, 3, \ldots\ldots\ldots$ *number of population* and D is the dimension of the space.

$$P_{sqa}^{cr}(i,j) = \begin{cases} P_{sqa}^{new}(i,j), & if\,(j_{rand} \le C_r)\,or\; j = j_{rand} \\ P_{sqa}^{old}(i,j), & if\,(j_{rand} > C_r)\,or\; j \ne j_{rand} \end{cases}, \quad j = 1, 2, 3 \ldots\ldots\ldots\ldots D \tag{18}$$

Now some squirrels update their position according to the squirrels at the neighborhood tree and other squirrels will follow their own current best at the normal tree. Therefore, the updated positions at other trees P_{sqoth} from these two facts can be expressed by (19) and (20), respectively.

$$P_{sqoth}^{new} = \begin{cases} P_{sqoth}^{old} + d_g G_c \times (P_{sqa}^{old} - P_{sqoth}^{old}), & r_2 \ge P_{pd} \\ Random\ position, & otherwise \end{cases} \tag{19}$$

$$P_{sqoth}^{new} = \begin{cases} P_{sqoth}^{old} + d_g G_c \times (P_{sqct}^{old} - P_{sqoth}^{old}), & r_3 \ge P_{pd} \\ Random\ position, & otherwise \end{cases} \tag{20}$$

Similarly, the crossover parameter is also applied to the squirrels at the other trees and expressed as in (21). Moreover, the convergence speed can be enhanced by introducing the average best positions to the current positions of the squirrel at the tree with food as (22).

$$P_{sqoth}^{cr}(i,j) = \begin{cases} P_{sqoth}^{new}(i,j), & if\,(j_{rand} \le C_r)\,or\; j = j_{rand} \\ P_{sqoth}^{old}(i,j), & if\,(j_{rand} > C_r)\,or\; j \ne j_{rand} \end{cases}, \quad j = 1, 2, 3 \ldots\ldots\ldots\ldots D \tag{21}$$

$$P_{sqct}^{new} = P_{sqct}^{old} + d_g G_c \times (P_{sqct}^{old} - P_{sqa}^{avg}) \tag{22}$$

The best value of the final positions of squirrels and their crossover rates are compared further with the previous positions in the process. For the MPPT of WT, the squirrel population is assigned as the rotor speed and chosen randomly. Further, based on the FF, sorting of the rotor speed population is performed iteratively to find out the global best solution. For a clear explanation, the DSSA optimization is presented in the form of pseudocode.

Pseudocode of DSSA algorithm for MPPT

```
Input maximum iterations N
Initialize rotor speed population
Initialize system parameters such as ub and lb
Calculate the fitness of squirrels (power for each assigned rotor value)
while (j < N) do
Sort the FF of all rotor speeds.
Calculate the current best positions of ω_wt in different locations.
Create the new position of ω_wt values using (17) following the updated
positions using (18).
Update new position of ω_wt values by comparing new positions obtained by
(19), (20), and (21).
Update new position of ω_wt by comparing the new positions obtained by (22)
and old position.
Update the population with the best position obtained so far
    j = j + 1
end while
return P_sqb
```

Grasshopper Optimization Based MPPT

The GOA is a new and metaheuristic algorithm that simulates the natural hunting and clustering activities of grasshoppers. Grasshoppers are bugs that are well for such a harmful bug that wreak havoc on agricultural production. The life-cycle is fragmented into two stages: maturity and nymph. Long range and rapid activities describe the adulthood stage but small steps and sluggish activities define the nymph stage, (Saremi, Mirjalili, and Lewis 2017). The diversification and intensification stages of GOA are characterized by adult and nymph movements.

Eq. (23) describes the position of ith grasshopper (ω_{wt}) in terms of social interaction factor, gravity force of a grasshopper $G_{i,}$ and wind advection. The gravity force of a grasshopper is given as the product of g_c, the gravitational constant and e_{grav}, a unit vector toward the center of the earth. Further, the product of the drift constant and is a unit vector in the wind direction is the wind advection. Here d_f represents the drift constant and e_{wind} is a unit vector in the wind direction. The social distance is the product of the Euclidean distance between the ith and the jth grasshopper and a unit vector from the ith

to the jth grasshopper, and k social forces in (24), where f is attraction strength and l is length scale.

$$X_i^{gh} = \sum_{\substack{j=1 \\ j\neq i}}^{N} k\left(\left|X_j - X_i\right|\right)\left(\frac{X_j - X_i}{d_{ij}}\right) - g_c \hat{e}_{grav} + d_f \hat{e}_{wind} \tag{23}$$

$$k(r) = f e^{\left(-\frac{r}{l}\right)} - e^{(-r)} \tag{24}$$

The above equation may not converge so (25) is used to enhance the performance. The upper limit and lower limit of dth dimension are extra parameters to enhance the best solution so far and are denoted by T_d.

$$X_i^{d^{th}} = c\left(\sum_{\substack{j=1 \\ j\neq i}}^{N} c.\frac{up_b - lw_b}{2} k\left(\left|X_j^{d^{th}} - X_i^{d^{th}}\right|\right)\left(\frac{X_j - X_i}{d_{ij}}\right)\right) + \hat{T}_d \tag{25}$$

Pseudocode of GOA algorithm for MPPT

```
Input maximum iterations N.
Initialize position of the cloud of grasshoppers (ωwt).
Initialize cmax and cmin.
Calculate the fitness of grasshoppers.
while (j < N) do
for each search agent do
Update current position of ωwt using Eq. (25).
end for
Update system parameters using Eqs. (26).
Calculate the fitness of all ωwt.
Update the global best position of each ωwt to obtain the maximum power.
    j = j + 1
end while
return Xα
```

Here, c is denoted as a comfort zone and can be expressed by (26), where c_{max} and c_{min} are maximum and minimum values of the comfort zone. t is the current iteration number and t_{max} is the maximum number of iterations.

$$c = c_{max} - t.\left(\frac{c_{max} - c_{min}}{t_{max}}\right) \tag{26}$$

Here, the rotor speed value is taken as the population of grasshoppers that are randomly assigned and to achieve the MPPT in WTS, rotor speed values are such varied iteratively that λ_{opt} can be achieved quickly and turbine power can be maximized.

Experimental Assesment

This section is prone to the performance investigation for the effectiveness of the discussed MPPT algorithms on 16 standard benchmark test functions. A thorough explanation for the benchmark test functions as well as initialization parameters are explored. Moreover, the performance analysis is done and the outcome is compared with various renowned metaheuristics algorithms on the various realistic engineering problems of optimization. The population size, total iterations, and dimension size are considered to be 20, 500, and 30, respectively for each algorithm. 30 independent runs per algorithm are performed for the statistical analysis on each benchmark test function for the evaluated algorithms.

Table 1. Unimodular benchmark functions

Function	*Dim*	*Range*	*fmin*
$f_1(x) = \sum_{i=1}^{n} x_i^2$	30	[-100, 100]	0
$f_2(x) = \sum_{i=1}^{n} \lvert x_i \rvert + \sum_{i=1}^{n} \lvert x_i \rvert$	30	[-10, 10]	0
$f_3(x) = \sum_{i=1}^{n} \left(\sum_{j-1}^{n} x_j\right)^2$	30	[-100, 100]	0
$f_4(x) = \max_i\{\lvert x_i \rvert, 1 \le i \le n\}$	30	[-100, 100]	0
$f_5(x) = \sum_{i=1}^{n-1}\left[100(x_{i+1} - x_i^2)^2 + (x_i - 1)^2\right]$	30	[-30, 30]	0
$f_6(x) = \sum_{i=1}^{n}(x_i + 0.5)^2$	30	[-100, 100]	0
$f_7(x) = \sum_{i=1}^{n} i x_i^4 + random[0,1]$	30	[-1.28, 1.28]	0

Table 2. Multimodular benchmark functions

Function	*Dim*	*Range*	f_{min}
$f_8(x) = \sum_{i=1}^{n} -x_i \sin(\sqrt{\lvert x_i \rvert})$	*30*	[-500, 500]	-418.989*5
$f_9(x) = \sum_{i=1}^{n} [x_i^2 - 10\cos(2\pi x_i) + 10]$	*30*	[5.12, 5.12]	0
$f_{10}(x) = \sum -20\exp\left(-0.2\sqrt{\frac{1}{n}\sum_{i=1}^{n} x_i^2}\right) - \exp\left(\frac{1}{n}\sum_{i=1}^{n}\cos(2\pi x_i)\right) + 20 + e$	*30*	[-32, 32]	0
$f_{11}(x) = \sum \frac{1}{4000}\sum_{i=1}^{n} x_i^2 - \prod_{i=1}^{n}\cos\left(\frac{x_i}{\sqrt{i}}\right) + 1$	*30*	[-600, 600]	0
$f_{12}(x) = \frac{\pi}{n}\{10\sin(\pi y_i) + \sum_{i=1}^{n}(y_i - 1)^2[1 + 10\sin^2(\pi y_i + 1)] + (y_n - 1)^2\} + \sum_{i=1}^{n} u(x_i, 10,100,4)$ $y_i = 1 + \frac{(x_i+1)}{4}$	*30*	[-50, 50]	0
$f_{13}(x) = 0.1\{\sin^2(3\pi x_i) + \sum_{i=1}^{n}(x_i - 1)^2[1 + \sin^2(3\pi x_i + 1)] + (x_n - 1)^2[1 + sin^2(2\pi x_n)] +$ $\sum_{i=1}^{n} u(x_i, 5,100,4)$	*30*	[-50, 50]	0
$f_{14}(x) = \left(\frac{1}{500} + \sum_{j=1}^{25}\frac{1}{j + \sum_{i=1}^{2}(x_i - a_{ij})^6}\right)^{-1}$	*2*	[-65, 65]	1
$f_{15}(x) = \sum_{i=1}^{11}\left[a_i - \frac{x_1(b_i^2 + b_i x_2)}{b_i^2 + b_i x_3 + x_4}\right]^2$	*4*	[-5, 5]	0.00030
$f_{16}(x) = [1 + (x_1 + x_2 + 1)^2(19 - 14x_1 + 3x_1^2 - 14x_2 + 6x_1x_2 + 3x_2^2)] \times [30 + (2x_1 - 3x_2)^2 \times$ $(18 - 32x_1 + 12x_1^2 + 48x_2 - 36x_1x_2 + 27x_2^2)]$	*2*	[-2, 2]	-3

To perform the statistical analysis, some values are needed to be assigned for each MPPT algorithms as initialization parameters. In case of WOA, $a \in [0, 2]$ and $a2 \in [-2, -1]$; in case of GOA, attraction strength is 0.5, length scale is 1.5, $[c_{min}, c_{max}] \in [0.00004, 1]$; in case of DSSA, gliding constant G_c is 0.5, crossover rate C_r is 0.3, random gliding distance d_g is 0.25 and predator presence probability P_{dp} is 0.3. In GWO and PSO-GWO algorithms, again $a \in [0, 2]$ and $a2 \in [-2, -1]$, c_1, c_2 and $c_3 = 0.5$. For the performance evaluation of all the discussed algorithms, some criteria are used as follows:

1. Standard deviation (SD): $SD = \sqrt{\frac{1}{iterations - 1} \sum_{i=1}^{iterations} (f_i - mean)^2}$

2. Mean: $Mean = \frac{\sum_{i=1}^{iterations} (f_i)}{iterations}$

The assessment for the optimum values can be done using the standard benchmark functions for both the functions i.e., unimodular and multi-modular. This assessment comprises exploration, exploitation, and convergence analysis of all the algorithms. In Table 1 and Table 2, all the tested functions are mentioned, where, D_{im} indicates the function dimension, *Range* is the search-space boundary, and f_{min} is the optimum value.

The exploitation for the unimodular functions and exploration for the multi-modular functions are conducted. The convergence statistic of all discussed algorithms has been compared in the graphs. As the iteration passes, the vacillations in DSSA reduce and it converges quicker than hybrid PSO-GWO and other algorithms as observed in Figures 6-8. Therefore, the DSSA provides improved convergence conduct and imposes a balance between exploitation and exploration in the space of iteration. As the population and iteration numbers are considered appropriate to the fastest algorithm, there may be a condition that the algorithm is not converged to the optimal value for all the functions while the focus is to express the comparative conduct among all the algorithms.

For the MPPT of modeled WTS, the convergence course might be restricted when executed with smaller population size and iterations, and the above conferred two parameters must give a faster real-world performance.

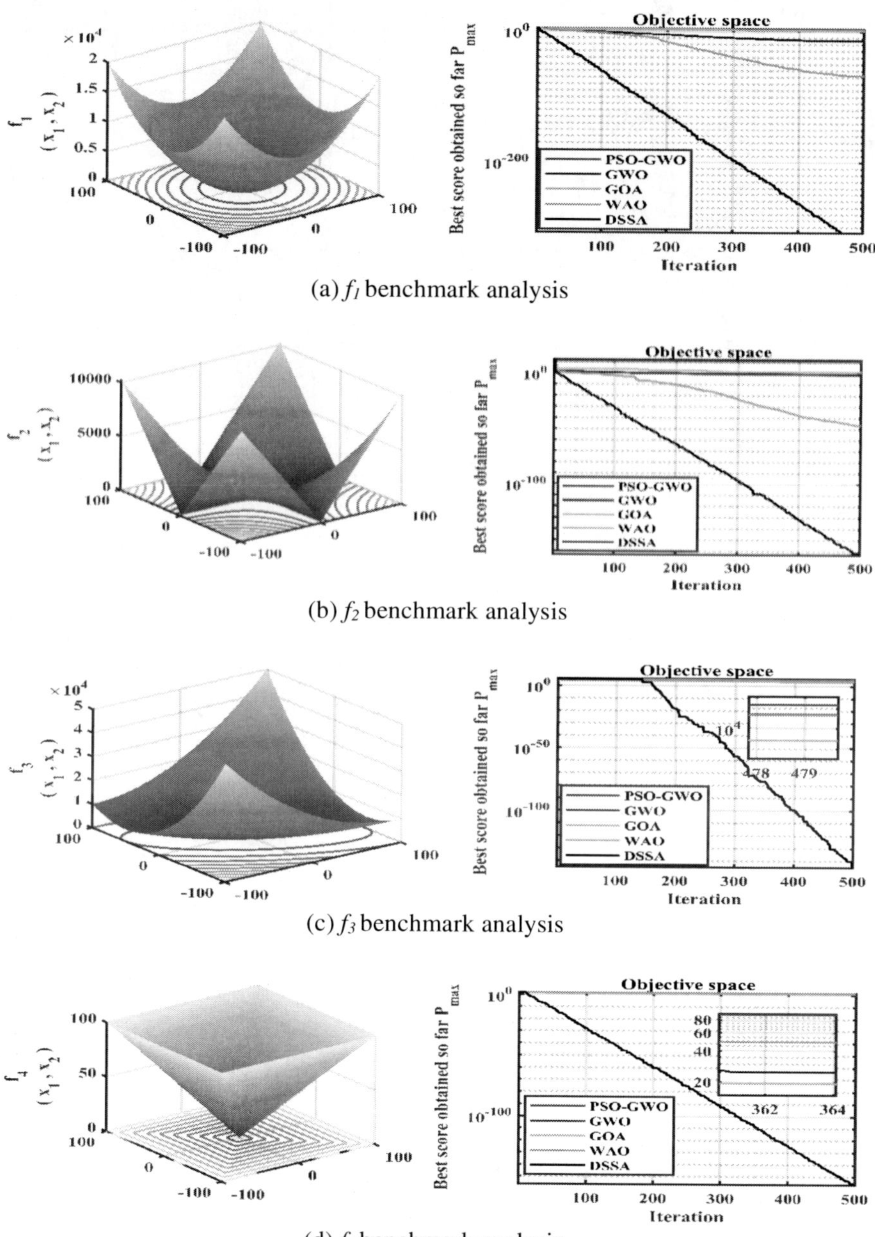

(a) f_1 benchmark analysis

(b) f_2 benchmark analysis

(c) f_3 benchmark analysis

(d) f_4 benchmark analysis

Figure 6. (Continued).

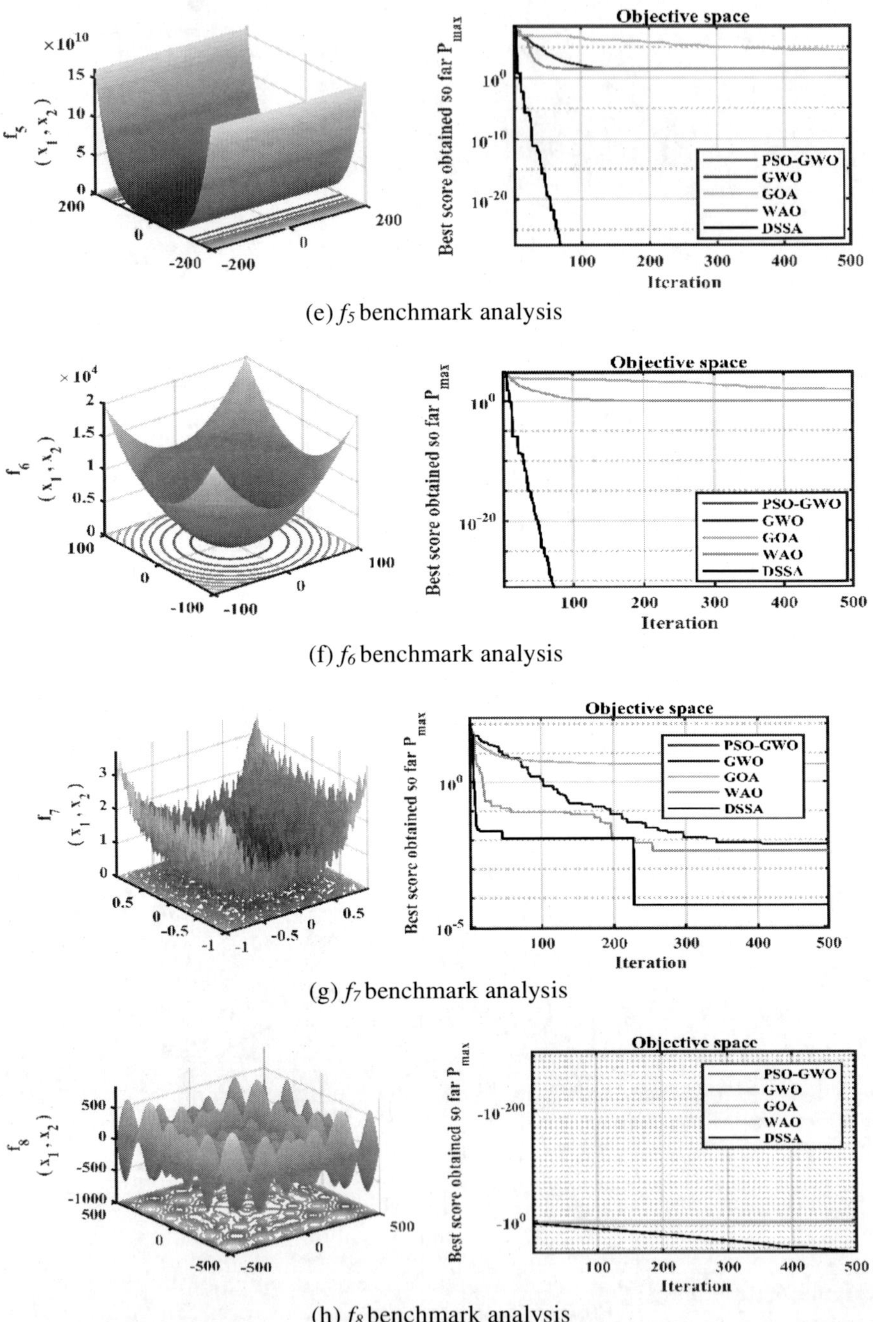

(e) f_5 benchmark analysis

(f) f_6 benchmark analysis

(g) f_7 benchmark analysis

(h) f_8 benchmark analysis

Figure 6. (Continued).

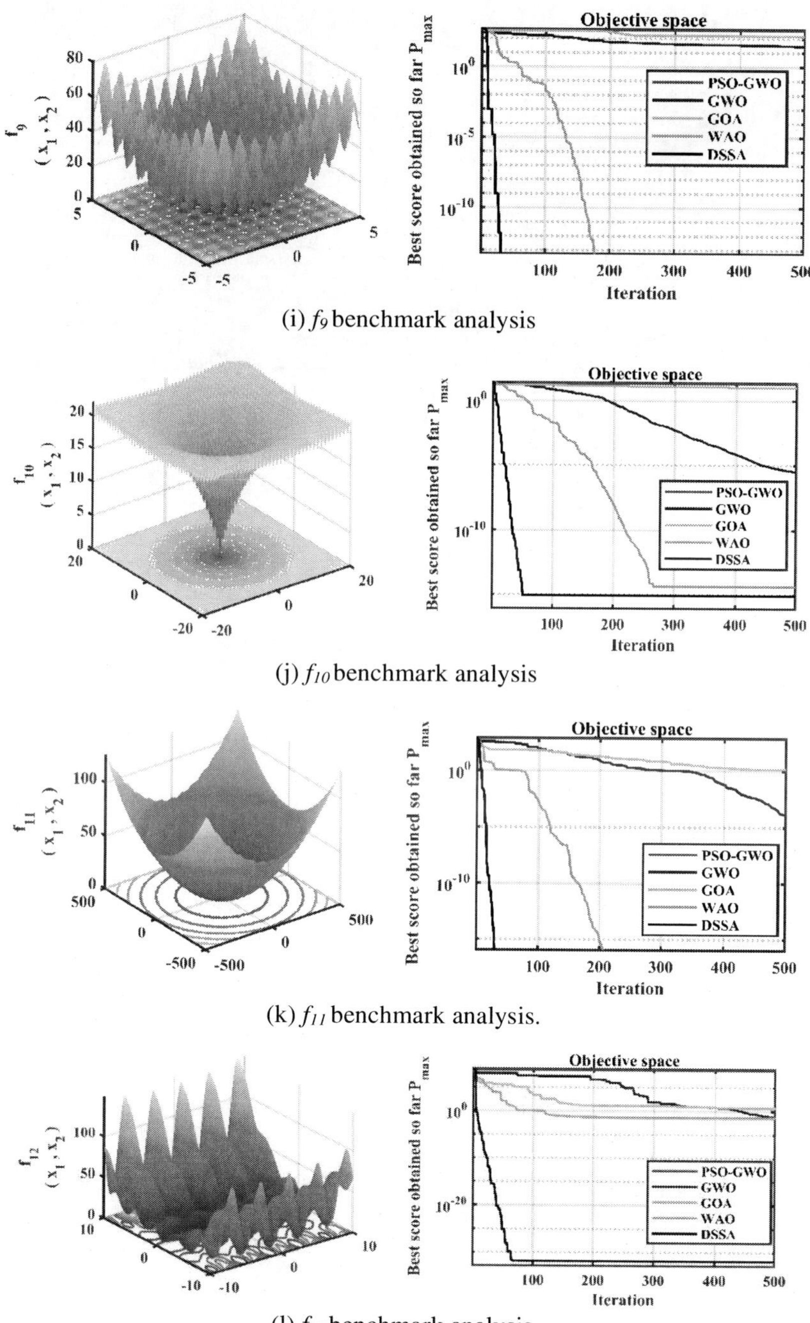

(i) f_9 benchmark analysis

(j) f_{10} benchmark analysis

(k) f_{11} benchmark analysis.

(l) f_{12} benchmark analysis

Figure 6. (Continued).

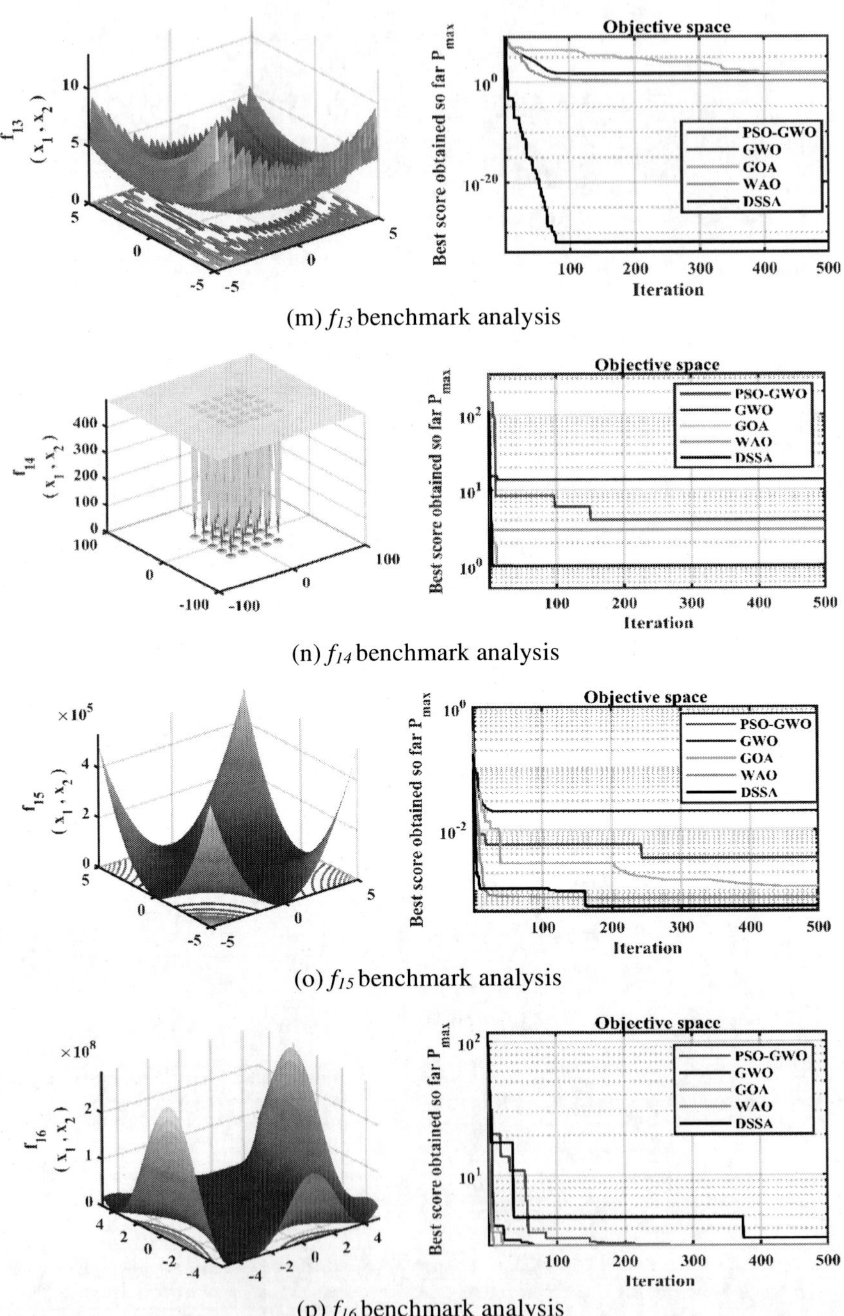

(m) f_{13} benchmark analysis

(n) f_{14} benchmark analysis

(o) f_{15} benchmark analysis

(p) f_{16} benchmark analysis

Figure 6 (a)-(p). Performance investigation for the discussed MPPT algorithms on standard benchmark test functions.

Table 3. The comparison of optimal value for the benchmark functions

Different Functions		PSO-GWO	GWO	GOA	WOA	DSSA
f_1	Mean	68051.446	9462.958	2557.464	8.86842E-09	5.46522E-59
	SD	6085.0710	12149.37499	1276.744572	0.000000017	0
	Median	67425.766	3511.276699	2076.810088	0.0000000007	0
f_2	Mean	5.46011E+13	11718383.13	64.79839899	8.08997E-08	6.84509E-31
	SD	1.6E+14	43914962.75	37.45084472	0.000000154	0
	Median	2.4E+12	8.723094456	73.34515498	0.000000012	0
f_3	Mean	126113.7592	23108.46193	8125.005406	113266.1503	4995.420731
	SD	25746.28652	23175.53512	2947.750029	31952.25928	21691.3796
	Median	123582.0726	12144.95455	6905.123175	110527.5746	0
f_4	Mean	88.73801087	26.97891281	23.71644295	70.11048475	1.03957E-23
	SD	4.135376184	18.2868111	4.182759387	21.57839028	0
	Median	89.67886069	20.32322704	23.8416006	80.31453418	0
f_5	Mean	256634572.9	11013089.4	621629.1624	28.81745692	5.741096731
	SD	49473890.41	25159957.36	409220.1706	0.033403738	11.48219346
	Median	249241898.7	25032.34099	640947.2248	28.8172097	0
f_6	Mean	69915.40732	5835.182289	2548.847203	2.448016517	0.002787584
	SD	6871.783449	12129.77626	1119.523872	0.74358216	0.012150798
	Median	69468.85243	88.93226242	2331.723024	2.454823331	0
f_7	Mean	131.5828835	1.372981385	11.17627826	0.023823159	0.0063361
	SD	27.88608164	2.677518549	4.494308596	0.025747538	0.005634855
	Median	135.3669971	0.326503794	9.947711769	0.013919425	0.004607598
f_8	Mean	-3309.081668	-6130.553259	-6412.253291	-9111.701698	-1.32433E+16
	SD	752.5163102	1093.919631	600.0433134	1475.94623	1.9888E+16
	Median	-3255.518556	-6385.971253	-6273.967858	-8868.446686	-4.37488E+15
f_9	Mean	450.9312323	251.0743133	189.0742545	7.052063215	11.70675907
	SD	28.30059565	108.1836365	50.07915781	29.14489815	38.37759748
	Median	456.6780944	303.8755536	186.2634529	0.000000008	0
f_{10}	Mean	19.96263802	9.978037578	12.30901396	4.58458E-06	8.88178E-16
	SD	0.001257152	6.193329829	2.534721383	0.000006726	0
	Median	19.96304577	10.35053787	12.13469774	0.000001416	0
f_{11}	Mean	633.8931108	30.26376553	20.22147589	2.01628E-08	0
	SD	76.12004445	82.10699221	8.884245148	0.000000069	0
	Median	622.1328014	6.147281574	18.77369125	0.0000000001	0
f_{12}	Mean	659229807.3	21471149.2	1271.566902	0.274517329	1.57054E-32
	SD	124011250.6	80138326.84	3789.270129	0.119450917	0
	Median	665817935.4	20.32619255	40.92107839	0.2645673617	0
f_{13}	Mean	1206334425	78301519.07	255075.2103	1.624684368	4.54474E-08
	SD	169094621.7	132787437.3	483339.4483	0.421559074	0.000000198
	Median	1196346223	651907.9113	63617.86562	1.674987469	0
f_{14}	Mean	4.689128907	6.709414957	11.77301894	4.42755457	1.595759069
	SD	2.549894929	5.034376687	8.143045639	4.071813733	2.541259954
	Median	4.358170918	5.57444639	10.76318067	2.982105239	0.998003837
f_{15}	Mean	0.011331539	0.005927938	0.015768368	0.002321679	0.000818983
	SD	0.008178624	0.008318342	0.014931449	0.004167366	0.000287519
	Median	0.008335426	0.001655717	0.014417313	0.000765756	0.000766493
f_{16}	Mean	3.414387468	7.062966623	24.60000095	5.904143753	6.507369953
	SD	0.408579902	17.65085629	34.78706818	8.458891112	3.418509177
	Median	3.325852378	3.000545999	3.000000003	3.000904901	6.032082746

The unimodular functions f_1-f_6 possess only single global optima, which is useful to evaluate the exploitation capacity of the discussed algorithms. It is apparent from Table 3 that the DSSA algorithm overperforms strongly as compared to the compared MPPT algorithms. The DSSA algorithm attains the best outcomes for most of the functions in all tested dimensions and, therefore, good exploitation can be obtained. On the other hand, multi-modular functions f_7-f_{16} possess multiple maxima and increase exponentially with the problem size. Hence, these functions are added to explore the better exploration capability. The DSSA algorithm continuously provides a better exploration analysis than the other studied algorithms. In some instants, the DSSA may provide the second-best outcomes as depicted in Table 3.

The convergence analysis is represented as the relation of FF values and iteration numbers for the studied algorithms. In this way, to reach the solution quickly by the algorithms is a foremost necessity to explore the graphical evaluation using the convergence rates. From Figure 6(a)-(p), the convergence responses of the DSSA are much faster than the other algorithms. Also, DSSA minimizes the FF value more than that of GWO, WOA, GOA, and PSO-GWO algorithms. It may be observed from Figure (a)-(p) that the convergence graph of the DSSA has not attained a constant minimized value. This is only to show a clearer picture of the convergence capability of other algorithms. Also, the iterations numbers are only considered as 500 for the analysis. In Figure 6(i)-(o), the successful convergence can be observed for the DSSA.

Result and Discussion

A comparative study of various MPPT algorithms for the WTS has been assessed for a 2 MW wind turbine in terms of tracking iterations (It_{track}), tracking efficiency (η_t), and obtained power (P_{obt}). Individual GWO, WAO, GAO, and two hybrid algorithms, PSO-GWO and DSSA MPPTs are implemented to effectively track the optimum value of the TSR, hence, the MPP of WT. The different cases have been examined and discussed below:

The efficacy of the discussed MPPT algorithms is examined, by performing a thorough simulation analysis under the effect of variable wind speeds. The MPPT scenarios are well-tested for the assigned populations of respective algorithms for each iteration and the evolutions are presented in Figure 8(a)-(e). Firstly, this evolution of the various population for assigned particles are assessed in terms of point of stabilization. Here, the point of stabilization is the point where the algorithm calculates the optimal value of the rotor speed corresponding to the optimal value of TSR. In the successive progression during the search for an optimal point, for a particular value of wind speed, particles are gathered at the possible best value that is expected to be obtained as shown in Figure 8(a)-(e). At the wind speed = 11 m/s, the particles have searched the stabilization point at 0.997 pu as the rated power of the WT is 2 MW at the rotor speed of 2.176 rad/s.

From Figure 8(a)-(e) it can be observed that DSSA and GOA based MPPT algorithms find the optimal point faster than that of WOA, GWO, and PSO-GWO MPPT algorithms, however, the GOA MPPT is stuck at local minima and fails to reach the optimal rotor speed, hence the maximum power. It can also be observed in Figure 7(b)-(c) clearly. On the other hand, DSSA MPPT tracks the MPP point effectively as compared to the other algorithm based MPPT method.

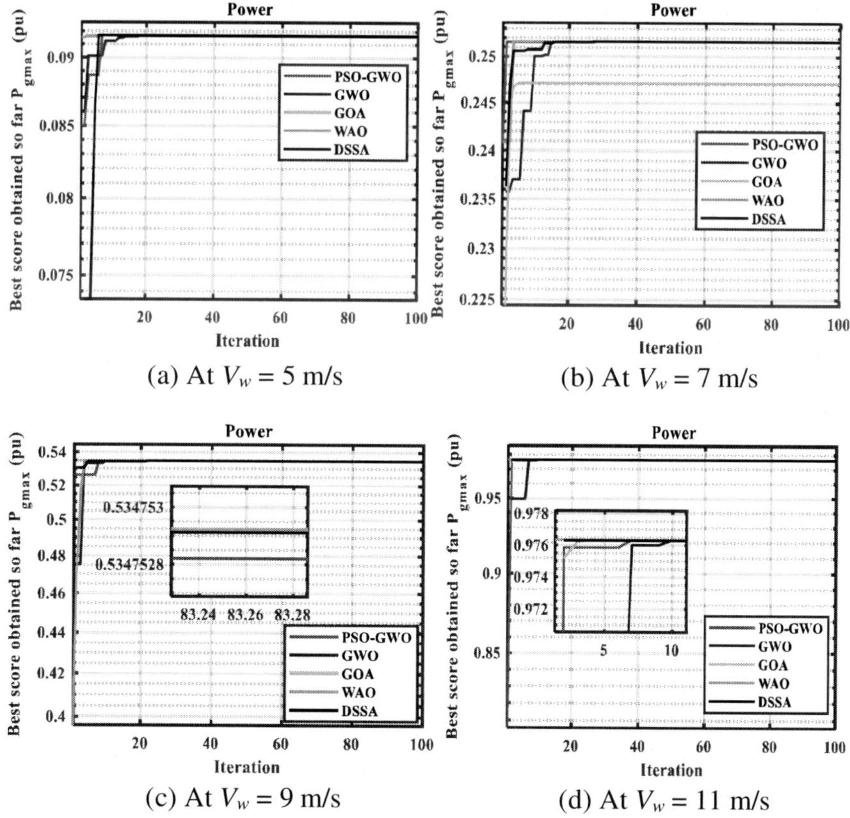

Figure 7 (a)-(d). Convergence graphs in terms of best score obtained using the MPPT algorithms.

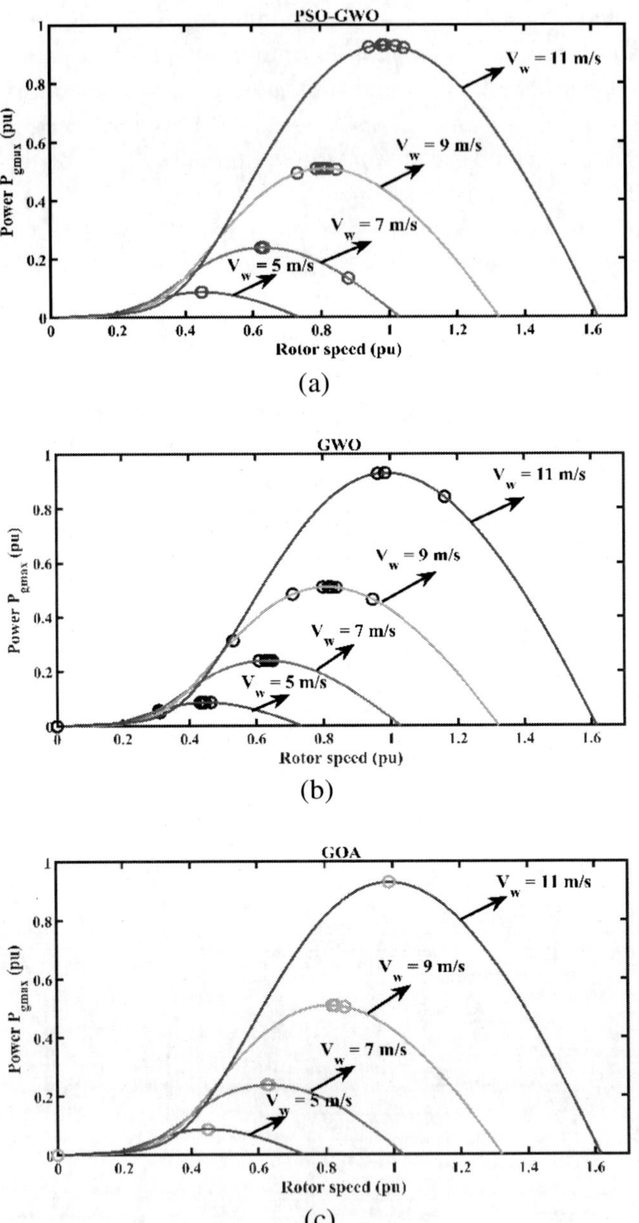

Figure 8. (Continued).

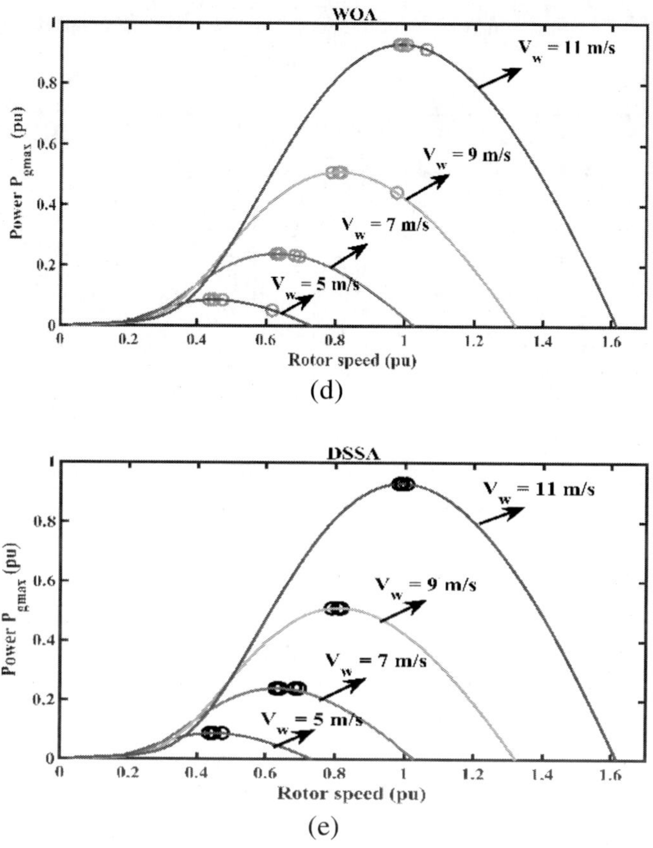

Figure 8 (a)-(e). Evolution of the MPPT algorithms during the search for an optimal point.

Table 4. Performance comparison of various MPPT algorithms at variable wind speed

MPPT algorithm	At V_w=11 m/s		At V_w = 9 m/s		At V_w = 7 m/s		At V_w = 5 m/s	
	I_{track}	η_t (%)	I_{track}	η_t (%)	I_{track}	η_t (%)	I_{track}	η_t (%)
PSO-GWO	7	99.99	7	99.77	2	99.99	6	99.99
GWO	9	99.99	9	98.40	13	99.99	8	99.99
GOA	3	99.99	7	65.21	6	84.05	6	99.99
WOA	2	99.99	3	99.98	3	99.99	2	99.99
DSSA	2	99.99	3	99.99	12	99.99	5	99.99

Table 5. Obtained power using various MPPT algorithms at variable wind speed

MPPT algorithm	P_{obt} (pu) at V_w=11 m/s	P_{obt} (pu) at V_w = 9 m/s	P_{obt} (pu) at V_w = 7 m/s	P_{obt} (pu) at V_w = 5 m/s
PSO-GWO	0.97634	0.5347528	0.25161	0.091693
GWO	0.97634	0.5347529	0.2516	0.091692
GOA	0.97635	0.53475292	0.24661	0.091693
WOA	0.97635	0.53475292	0.25161	0.091693
DSSA	0.97635	0.5347529	0.25156	0.091693

Overall performance comparison among the various studied MPPT method is assessed in Table 4 and Table 5. Table 4 compares the MPPT performances in terms of the number of iterations taken to converge at the optimal point and tracking efficiency at the different wind speed conditions. It can be observed that DSSA MPPT achieves the MPP in minimum iterations, e.g., at V_w = 11 m/s, DSSA takes 2 iterations to reach the MPP, whereas, PSO-GWO and WOA MPPTs require 10 and 2 iterations, respectively to reach MPP. However, the convergence time depends upon the dynamics of the WTS at various wind speeds. All MPPT algorithms achieve the optimal point except the GOA MPPT as can be observed in Table 4. At some value of the wind speed, GOA MPPT gets stuck at local minima and fails to track the optimal power. Table 5 explains the obtained turbine power by each MPPT algorithm. The foremost strength shown by the metaheuristic-based DSSA MPPT algorithms is the minimized oscillations around the MPP and its convergence speed.

Conclusion

In this work, various metaheuristic-based MPPT algorithms like WOA, GOA, GWO, hybrid PSO-GWO, and DSA are developed and implemented to achieve the maximum efficiency of a large-scale WTS. Each algorithm is well-tested on the standard unimodular and multi-modular benchmark functions before applying them to the MPPT loop. Exploitation analysis for unimodular and exploration analysis for multi-modular reveal that DSSA has the competency to achieve the best results in terms of mean, standard deviation, and median. Further, based on the aforementioned analysis, each algorithm is applied for the MPPT of WTS. The performance of each MPPT method is compared in terms of the number of iterations taken for convergence, tracking efficiency, and tracked power. It is found that the DSSA MPPT method provides competing results as compared to other studied MPPT methods. The analysis presented in this work may provide an important base for the involvement of the grid stabilization integrated with the present WTS and back-to-back PWM converters.

References

Ali, Youssef Ait, and Mohammed Ouassaid. 2019. "Sensorless MPPT Controller Using Particle Swarm and Grey Wolf Optimization for Wind Turbines." *Proceedings of 2019 7th International Renewable and Sustainable Energy Conference, IRSEC 2019* 2 (3). https://doi.org/10.1109/IRSEC48032.2019.9078151.

Fathy, Ahmed, Mohammad Ali Abdelkareem, A. G. Olabi, and Hegazy Rezk. 2021. "A Novel Strategy Based on Salp Swarm Algorithm for Extracting the Maximum Power of Proton Exchange Membrane Fuel Cell." *International Journal of Hydrogen Energy* 46 (8): 6087–99. https://doi.org/10.1016/j.ijhydene.2020.02.165.

Fathy, Ahmed, and Osama El-Baksawi. 2019. "Grasshopper Optimization Algorithm for Extracting Maximum Power from Wind Turbine Installed in Al-Jouf Region." *Journal of Renewable and Sustainable Energy* 11 (3). https://doi.org/10.1063/1.5085167.

Jain, Mohit, Vijander Singh, and Asha Rani. 2019. "A Novel Nature-Inspired Algorithm for Optimization: Squirrel Search Algorithm." *Swarm and Evolutionary Computation* 44 (November 2017): 148–75. https://doi.org/10.1016/j.swevo.2018.02.013.

Jena, Bibekananda, Manoj Kumar Naik, Aneesh Wunnava, and Rutuparna Panda. 2021. *A Differential Squirrel Search Algorithm. Lecture Notes in Networks and Systems*. Vol. 202 LNNS. Springer Singapore. https://doi.org/10.1007/978-981-16-0695-3_15.

Kumar, Dipesh, and Kalyan Chatterjee. 2016. "A Review of Conventional and Advanced MPPT Algorithms for Wind Energy Systems." *Renewable and Sustainable Energy Reviews* 55: 957–70. https://doi.org/10.1016/j.rser.2015.11.013.

Kurian, Geevarghese Mathew, Aruna Jeyanthy and D.Devaraj. 2022. "FPGA implementation of FLC-MPPT for harmonics reduction in sustainable photovoltaic system." *Sustainable Energy Technologies and Assessments* 52: 102192. https://doi.org/10.1016/j.seta.2022.102192.

Lee, Joonmin, and Young Seok Kim. 2016. "Sensorless Fuzzy-Logic-Based Maximum Power Point Tracking Control for a Small-Scale Wind Power Generation Systems with a Switchedmode Rectifier." *IET Renewable Power Generation* 10 (2): 194–202. https://doi.org/10.1049/iet-rpg.2015.0250.

Mallipeddi, R., P. N. Suganthan, Q. K. Pan, and M. F. Tasgetiren. 2011. "Differential Evolution Algorithm with Ensemble of Parameters and Mutation Strategies." *Applied Soft Computing Journal* 11 (2): 1679–96. https://doi.org/10.1016/j.asoc.2010.04.024.

Mirjalili, Seyedali, and Andrew Lewis. 2016. "The Whale Optimization Algorithm." *Advances in Engineering Software* 95: 51–67. https://doi.org/10.1016/j.advengsoft.2016.01.008.

Mirjalili, Seyedali, Seyed Mohammad Mirjalili, and Andrew Lewis. 2014. "Grey Wolf Optimizer." *Advances in Engineering Software* 69: 46–61. https://doi.org/10.1016/j.advengsoft.2013.12.007.

Mokhtari, Yacine, and Djamila Rekioua. 2018. "High Performance of Maximum Power Point Tracking Using Ant Colony Algorithm in Wind Turbine." *Renewable Energy* 126: 1055–63. https://doi.org/10.1016/j.renene.2018.03.049.

Mousa, Hossam H.H., Abdel Raheem Youssef, and Essam E.M. Mohamed. 2019. "Variable Step Size P&O MPPT Algorithm for Optimal Power Extraction of Multi-Phase PMSG Based Wind Generation System." *International Journal of Electrical Power and Energy Systems* 108 (December 2018): 218–31. https://doi.org/10.1016/j.ijepes.2018.12.044.

Pathak, Diwaker, Sachin Bhati, and Prerna Gaur. 2020. "Fractional-Order Nonlinear PID Controller Based Maximum Power Extraction Method for a Direct-Driven Wind Energy System." *International Transactions on Electrical Energy Systems* 30 (12): 1–27. https://doi.org/10.1002/2050-7038.12641.

Pathak, Diwaker, and Prerna Gaur. 2019. "A Fractional Order Fuzzy-Proportional-Integral-Derivative Based Pitch Angle Controller for a Direct-Drive Wind Energy System." *Computers and Electrical Engineering* 78: 420–36. https://doi.org/10.1016/j.compeleceng.2019.07.021.

Pathak, Diwaker, and Prerna Gaur. 2020. "A Novel Fractional-Order Nonlinear PID-Based Pitch Angle Control Strategy for a PMSG-Based Wind Energy System." *International Transactions on Electrical Energy Systems* 30 (12): 1–22. https://doi.org/10.1002/2050-7038.12671.

Priyadarshi, Neeraj, Vigna K. Ramachandaramurthy, Sanjeevikumar Padmanaban, and Farooque Azam. 2019. "An Ant Colony Optimized Mppt for Standalone Hybrid Pv-Wind Power System with Single Cuk Converter." *Energies* 12 (1). https://doi.org/10.3390/en12010167.

Rezk, Hegazy, Ahmed Fathy, and Almoataz Y. Abdelaziz. 2017. "A Comparison of Different Global MPPT Techniques Based on Meta-Heuristic Algorithms for Photovoltaic System Subjected to Partial Shading Conditions." *Renewable and Sustainable Energy Reviews* 74 (August 2016): 377–86. https://doi.org/10.1016/j.rser.2017.02.051.

Sagar, Gautam, Diwaker Pathak, Prerna Gaur, and Vatsal Jain. 2020. "A Su Do Ku Puzzle Based Shade Dispersion for Maximum Power Enhancement of Partially Shaded Hybrid Bridge-Link-Total-Cross-Tied PV Array." *Solar Energy* 204 (March): 161–80. https://doi.org/10.1016/j.solener.2020.04.054.

Saremi, Shahrzad, Seyedali Mirjalili, and Andrew Lewis. 2017. "Grasshopper Optimisation Algorithm: Theory and Application." *Advances in Engineering Software* 105: 30–47. https://doi.org/10.1016/j.advengsoft.2017.01.004.

Shaheen, Mohamed A.M., Hany M. Hasanien, and Abdulaziz Alkuhayli. 2021. "A Novel Hybrid GWO-PSO Optimization Technique for Optimal Reactive Power Dispatch Problem Solution." *Ain Shams Engineering Journal* 12 (1): 621–30. https://doi.org/10.1016/j.asej.2020.07.011.

Sitharthan, R., Madurakavi Karthikeyan, D. Shanmuga Sundar, and S. Rajasekaran. 2020. "Adaptive Hybrid Intelligent MPPT Controller to Approximate Effectual Wind Speed and Optimal Rotor Speed of Variable Speed Wind Turbine." *ISA Transactions* v96: 479–89. https://doi.org/10.1016/j.isatra.2019.05.029.

Wang, Xiaolong, Yuling He, Haipeng Wang, Aijun Hu and Xiong Zhang. 2022. "A novel hybrid approach for damage identification of wind turbine bearing under variable speed condition." *Mechanism and Machine Theory* 169: 104629. https://doi.org/10.1016/j.mechmachtheory.2021.104629.

Zouheyr, Dekali, Baghli Lotfia and Boumediene Abdelmadjid. 2021. "Improved hardware implementation of a TSR based MPPT algorithm for a low cost connected wind turbine emulator under unbalanced wind speeds." *Energy* 232: 121039. https://doi.org/10.1016/j.energy.2021.121039.

Chapter 9

Wind Power Prediction Using Hybrid Soft Computing Models

Pavan Kumar Singh[1], Parvez Ahmad[2], Nitin Singh[2], Niraj Kumar Choudhary[2] and Richa Negi[2]

[1]Shambhunath Institute of Engineering and Technology Allahabad, Prayagraj, Uttar Pradesh, India

[2]Motilal Nehru National Institute of Technology Allahabad, Prayagraj, Uttar Pradesh, India

Abstract

Wind energy integration into smart grids has become popular recently because of its availability in abundance. The main problem in integrating wind power in smart grids is the wind's irregular and unpredictable nature. Therefore, accurate forecasting methods are pivotal in planning and executing wind energy integration. This chapter presents the two-hybrid soft computing-based techniques for wind power forecasting. Combining wavelets with an optimized ANFIS model is the first proposed method. The second method combines wavelets and dynamic recurrent neural networks, NAR and NARX. For a year, average hourly wind power data is collected through a wind power plant in Agar, USA. Wavelet transform is employed to decompose the data before applying it to the optimized ANFIS model and NAR and NARX networks to predict wind power in each case. After that, the output sub-series is combined to obtain the final predicted wind power. The proposed hybrid Wavelet Neuro-Fuzzy-DPSO approach has been compared to other intelligent techniques, and the WT-NARX approach was compared to the WT-NAR and persistence methods. In each case, the forecast results of the hybrid models revealed the best performance compared to the other models.

Keywords: wind power prediction, adaptive network-based fuzzy inference system (ANFIS), dynamic recurrent neural network, nonlinear autoregressive network (NAR), dynamic particle swarm optimization

(DPSO), nonlinear autoregressive network with exogenous input network (NARX), wavelet transform (WT)

Introduction

In real-world problems, uncertainty is a significant challenge that leads to incomplete and imprecise information about the input data. The large-scale integration of wind power in a smart grid is challenging for system stability and reliability due to its chaotic nature and intermittency. Wind energy generation depends not only on the wind speed but is also affected by the obstacle and terrain. The randomness of wind is significant due to its variation with height. The electricity produced by wind energy should be consumed instantaneously. As a result, the synchronized timing of load and wind patterns decides the profitable value of wind power. That is why wind energy-based generators cannot be scheduled to meet variable load [1]. Integrating large-scale wind energy deployment can create positive social and environmental impacts. However, wind energy's stochastic and unpredictable characteristic is a critical concern that is slowing the industry's progress.

The ANN-based forecasting model has become popular recently as it produces promising results [2]. They can quickly adapt to the continuously changing forecasting conditions/environment through self-learning. However, the suitability of the ANN model is determined using a trial-and-error process, leading to over fitting the model and unacceptable forecasting results [3-4]. The major problem with the parametric models is that the exact model parameters need to be defined, which is a cumbersome process. The forecasting accuracy depends upon the value of the chosen parameters, and the model coefficients must be re-calculated for a slight change in the input signal. The artificial neural network (ANN) model has its own limitations [5].

The model's forecasting accuracy, which is critical for prediction, is improved by combining multiple soft computing approaches. Two-hybrid intelligent models are suggested in this chapter. To overcome the issues of ANN and ARIMA models for time series prediction, the first model incorporates the wavelet transform (WT) with ANFIS. The ANFIS model parameters are optimized using the DPSO algorithm to improve the model's execution even more.

The second hybrid model uses *the dynamic* (*recurrent*) *Neural Networks*, which have been used to analyze the time-dependent wind power series data. These models have successfully been implemented in the literature to predict the time series. Analysis and modeling of the time-dependent processes are most appropriately achieved using DNNs [6]. DNNs are flexible because they use various processing features, including feedback and delay, to convey dynamic behavior. The memory order specifies the steps by which the signal will be delayed. The complete system comprises the network topology (architecture) and the control system for the signal time delay.

The rest of the sections of this chapter are prepared as follows. Section 2.1 includes the wavelet decomposition of time series, mother wavelet selection, and decomposition level of time series. Section 2.2 describes ANFIS architecture, and two widely used dynamic recurrent neural networks (NAR and NARX) are discussed in section 2.3. Section 2.4 briefs about dynamic particle swarm optimization (DPSO). Sections 3 and 4 describe the suggested WT-ANFIS-DPSO, and WT-NAR/NARX hybrid approaches. Finally, the conclusion of the proposed hybrid methods has been presented in Section 5.

Wind Power Prediction Techniques

The suggested method uses a hybrid intelligent technique that combines wavelets and ANFIS to estimate wind power. The DPSO technique optimizes the fuzzy inference system's membership functions (MFs) parameters to achieve minimal forecasting error. In addition, the subseries of wind power obtained using wavelet transform is predicted using dynamic recurrent neural networks (NAR and NARX) hybrid models.

Wavelet Transform (WT)

The WT has recently become popular as it can detect outliers in a series. For wind power forecasting, the wavelet transform has been frequently employed. The time series has been assumed to be stationary by most studies. On the other hand, others believe it is a white noise deterministic function. The de-noising ability of WT improves the features of ill-behaved series, allowing for more accurate predictions. The non-stationary data having non-constant mean and autocorrelation are analyzed using the WT. The WT mathematically splits the time series into two constituent series: approximate (representing high-frequency part] and detail (representing low-frequency part). The Fourier transform is used to split the signal into a weighted sum of cosine and sine functions. On the contrary, WT separates the original signal into wavelets that are more flexible and localized in terms of time and frequency [7-9].

The mother wavelet is characterized as a wavelet prototype in WT. The mother wavelet's high-frequency part evaluates the signal's temporal characteristics. The mother wavelet's low-frequency component does the frequency analysis of the original signal. The continuous wavelet transforms (CWT) is used with the continuous signals, whereas the discrete wavelet transform (DWT) is used with discrete-time signals. Mallat proposed a multi resolution signal decomposition that divides the time signal into several resolution signals. In order to restore the original signal, the inverse WT [10] is used.

A multiresolution pyramidal decomposition technique is used to implement the WT. The high and low-frequency parts of the actual signal S(n) are extracted utilizing high and low pass filters. As a result, the signal's high-frequency part, also known as the detailed component, and low-frequency part, also known as the approximation component, are determined. The Haar, Meyer, Coiflet, Daubechies, and Morlet are more commonly employed as mother wavelets among the different mother wavelets found in the literature. Figure 1 shows the signal decomposition with three decomposition levels using the wavelet transform.

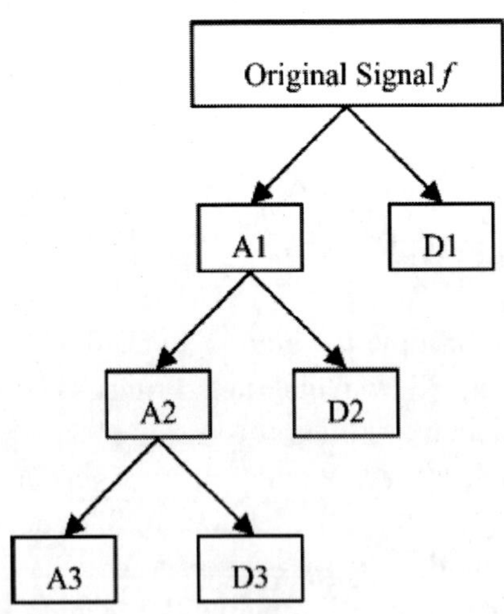

Figure 1. Multilevel decomposition of the signal.

The wavelets convert the input series into many sub-series, which improves the behaviour of the original series, leading to more accurate predictions. The mother wavelet selection and level of decomposition are the two essential aspects that need to be looked upon before applying the wavelet decomposition to any time series. The mother wavelet is chosen based on the original signal's resemblance to the approximated signal. The wavelet transform is chosen by comparing the mother wavelet's features to the characteristics of the original signal. Daubechies and Haar wavelets are from the ortho-normal wavelets and are most appropriate for the non-stationary time series. Furthermore, Haar and Daubechies are discrete wavelets transformed with a compact or narrow window, and hence they are more suitable for local signal analysis [11].

The level of decomposition is selected by close examination of the approximate part and the detailed part's similarity in shape compared with the original signal. The time-series data from the wind park is decomposed using the Daubechies mother wavelet of order four (i.e., db4) and three-level decomposition in the proposed study. The Daubechies mother wavelet analyses average hourly data of wind speed, wind power, air temperature, and air pressure. Smoothness grows with the order of the function in the Daubechies family of mother wavelets, as does the support interval, causing the prediction to deteriorate. Therefore, it is recommended to have lower-order wavelet functions [12-13].

The wind power series variables are decomposed using Daubechies wavelet. The order of decomposition is kept as four, and the level of decomposition is kept as three as it significantly characterizes the series. The equation (1) shows the wavelet transforms W(a,b) of the signal f(x), transformed using the mother wavelet Ø(x). The variable 'a' is called the scale factor, responsible for expanding the wavelet. The variable 'b' is called the translation factor, responsible for its central position [14].

$$W(a,b) = \frac{1}{\sqrt{a}} \int_{-\infty}^{+\infty} f(x) \emptyset\left(\frac{x-b}{a}\right) dx \tag{1}$$

Adaptive Network-Based Fuzzy Inference System (ANFIS)

ANFIS is an adaptive multilayer feed forward neural network mainly used to forecast nonlinear series. The ANN and FINS models are combined in the ANFIS model. The ANN is used to declare the ANFIS model's fuzzy inference functions. The FIS expresses fuzzy logic on a scale of zero to one using conventional fuzzy IF-THEN rules. The IF part describes the antecedent rules, and the consequent or conclusion rules are illustrated in the THEN portion. The database employed in the fuzzy rules describes the membership functions (MFs) of fuzzy sets. The decision-making unit performs the inference operation on the fuzzy rules. The membership functions turn the crisp value into a fuzzy value in the fuzzification process. Finally, the Defuzzification process converts back the fuzzy value to the crisp value. Each layer's nodes are represented by node functions [15-17]. Every node is an adaptive node in layer one that yields the output as shown in equations (2) and (3).

$$O_{1,i} = \mu_{Ai}(x), i = 1,2 \tag{2}$$

$$O_{1,i} = \mu_{Bi-2}(y), i = 3,4 \tag{3}$$

where $O_{1,i}$ denotes the fuzzy set membership grade. The antecedent membership functions, i.e., μ_{Ai}, μ_{Bi-2} are usually bell-shaped functions with maximum value as one and minimum as zero, as shown in equation (4).

$$\mu_{Ai}(x) = \frac{1}{1+\left|\frac{x-r_i}{p_i}\right|^{2qi}} \tag{4}$$

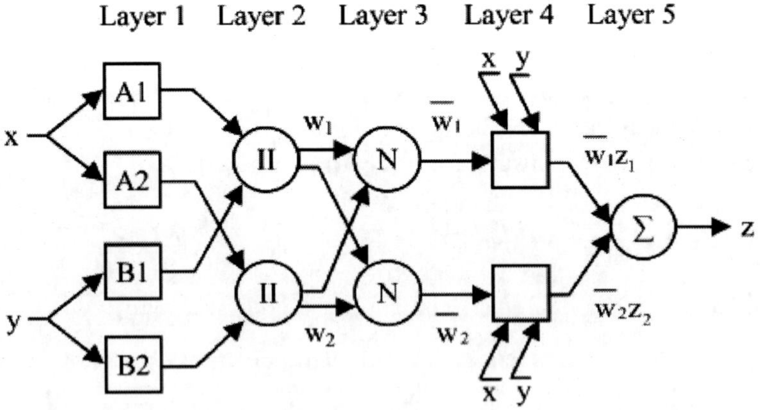

Figure 2. The architecture of the ANFIS model.

A standard rule expresses each node in layer two, i.e., the rule layer. It is predetermined, as shown in equation (5).

$$O_{2,i} = w_i = \mu_{Ai}(x)\mu_{bi}(y) \quad (5)$$

The output of each node is a normalized firing strength of each rule in layer three, i.e., the normalizing layer, as indicated in equation (6).

$$\overline{w}_i = \frac{w_i}{w_1+w_2} = O_{3,i} \quad (6)$$

The weighted outcomes of each rule are calculated at layer four at each node. As stated in equation (7), it correlates to the rules of all nodes for obtaining the output (7).

$$O_{4,i} = \overline{w}_i z_i = \overline{w}_i(a_i x + b_i y + c_i) \quad (7)$$

In layer five, i.e., the summation layer, a single node adds all the outputs of layer four, as shown in equation (8) [18].

$$O_{5,i} = \sum_i \overline{w}_i z_i = \frac{\sum_i w_i z_i}{\sum_i w_i} \quad (8)$$

Dynamic Recurrent Neural Networks (DNNs)

The commonly used forecasting methods are well-suited to linear time series only. In contrast, nonlinear methods are required for nonlinear forecasting of series like wind power and wind speed. One of the nonlinear techniques deploys an artificial neural network architecture where the input data varies nonlinearly, and after proper training, the network can predict future values. The proposed work uses nonlinear autoregressive with exogenous input (NARX) and nonlinear autoregressive (NAR) to forecast the wind power series.

The single variable is used in the univariate modeling, utilising the historical values of wind power to forecast future values. On the other hand, the multivariate model, i.e., NARX, requires two or more input variables for modeling and prediction. The topologies of NAR and NARX are shown in Figures 3 and 4, respectively. The Levenberg-Marquardt (LM) algorithm is utilized for training both models. The numbers of input lags are selected using ACF and PACF [19]. The cross-correlation functions (CCFs) between the residuals of output and each input time series are specified by the number of lags at which maximum correlation occurs. After selecting the maximum lag kmax, the values at lags 1 to kmax are utilized as model inputs. The need for forecasting steps (periods) determines the number of output lags [20]. The NARX model improves the time series forecasts due to external (exogenous) information fed as input [21]. The neural network toolbox in MATLAB is used to simulate both the dynamic neural network models [22].

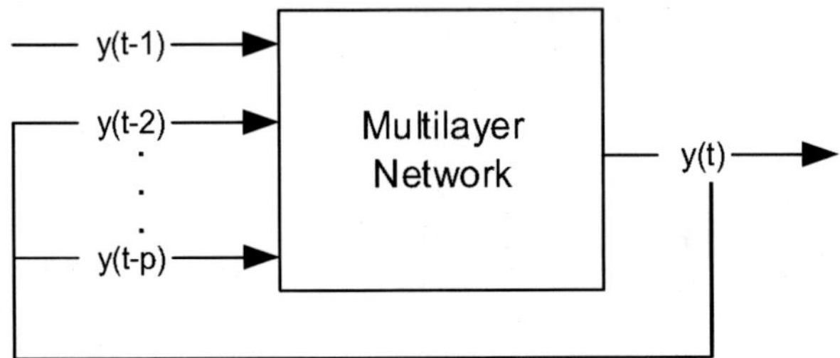

Figure 3. Topology of a NAR network.

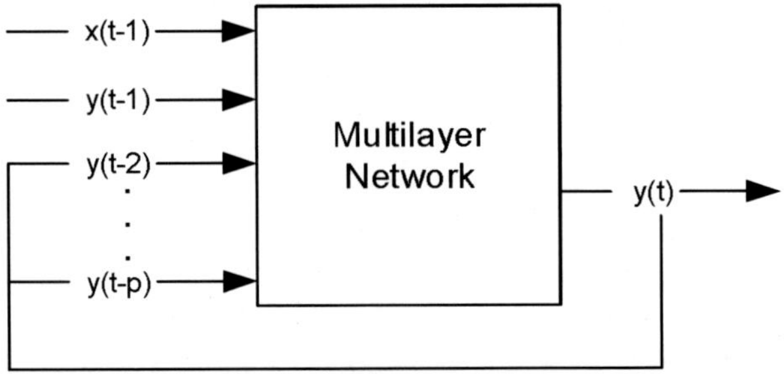

Figure 4. Topology of a NARX network.

NAR Neural Network

The NAR, a recurrent neural network, is suitable for assessing the future input variables using univariate time series delays. When employed for time series forecasting, the recurrent type neural network represents a discrete, nonlinear, autoregressive model. The forecasting done using the NAR network utilizes the historical values of the time series. The predicted value serves as an input in a re-feeding mechanism for a new prediction. The NAR network is trained using an open-loop architecture similar to a three-layer feed forward structure. The actual target values are used as feedback to build and train the network, thus ensuring better training accuracy. The network is converted to a closed-loop after training, and the predicted values are feedback to the network. The prediction of the NAR network is a function of preceding observation values, as shown in equation (9).

$$y(t) = f\,(y(t\text{-}1), y(t\text{-}2), \ldots\ldots, y(t\text{-}p)) \tag{9}$$

y(t) and p denote the series value at time t and the number of feedback delays, respectively.

Equation (9) illustrates how the value y(t) of a data series y at time t is predicted by a NAR network utilizing p previous time series values. The function f(.) is not identified earlier and approximated by neural network training with optimization of weights and bias of neurons in a neural network. The p features of time series y(t-1), y(t-2), . . , y(t-p) are known as feedback delays, as shown in Figure 3. The functions of the hidden layer and the number of neurons are highly significant from several successful applications of neural networks. The neurons in the hidden layer perform the feature detection and capture any existing pattern in the data and complex input-output nonlinear mapping of variables. Ideally, neural networks can estimate any nonlinear function with the desired accuracy with just one hidden layer.

The trial-and-error method optimizes the total number of hidden layers and neurons per layer and gets the network topology for best performance. However, excess neurons can make the system complicated. In contrast, the insufficient number of neurons reduces its generalization capability and computational efficiency. The Levenberg-Marquardt backpropagation (LMBP) is generally the learning rule applied to the NAR networks because it is the fastest algorithm. In addition, LMBP algorithms do not require computing the Hessian matrix to approximate the second-order derivative, which increases the training speed [23-24].

NARX Neural Network

There is a significant correlation between the modeled time series and added exogenous data in the numerous real applications. It has been shown that the high dependency of the wind power plant's production on the weather environment is a common characteristic of renewable energy. Therefore, to provide an accurate forecast, the knowledge or data integration concerning weather could benefit the time series modeling compared to any single variable related to wind power. The NARX architecture is used if the model has more associated variables. NARX models predict wind power series y(t) using p historical values of the time series y(t) and an extra exogenous time series x(t) that may be single or multidimensional. The behaviour of the

NARX neural network model for time series prediction can be expressed as equation (10).

$$y(t) = h(x(t-1), x(t-2), \ldots\ldots, x(t-k), y(t-1), y(t-2), \ldots\ldots, y(t-p)) \quad (10)$$

The NARX model uses the previous output and exogenous input to predict the values of a time series. NARX model utilizes past values of wind power time series y(t) as one input and wind speed data at x(t) as other exogenous input and gives one step ahead output y(t) concerning to value of wind power. The NARX architecture, as shown in Figure 4, is identical to the NAR network, except that their inputs are distinct. Hence, external data is considered by output y(t), as shown in equation (10). For training the NARX model, LMBP is used as the learning algorithm.

In neural networks, selecting input variables is necessary because the chosen features highly influence prediction accuracy. The ANN input lags are traditionally determined based on the trial and error method. Several neural networks should be created and checked with various numbers of input lags. After that, the superior performance networks would be chosen as the forecasting model. This time-consuming method can be avoided by applying two statistical measures, ACF and PACF, to select input lags of variables. The analysis of ACF and PACF is performed for sub-series of signals after decomposition. The number of lags significantly correlates to future wind power and wind speed values, as shown by the ACF and PACF plots so that they can be used as the neural network's input. The positive peaks above the confidence limit suggest the NARX model's optimal lag. A complex error minimization algorithm is needed for the large datasets to determine optimal lag [25-27].

Dynamic Particle Swarm Optimization (DPSO)

The PSO influenced by the population behaviour of the swarm is used to determine the optimization problem's solution. The PSO algorithm changes the particle's velocity using the global best [24]. However, the particle's position also gets influenced by the global best, leading to faster convergence. Hence, it may stop at local optima and not reach global optima, particularly in the multimodal problem [25]. A new variant of PSO was presented in [26], which achieves both objectives, i.e., the convergence speed and avoidance of the local minima of the population-based algorithm. Because of contradiction in both objectives, they are otherwise difficult to achieve.

DPSO is a commonly used PSO variant parameter optimization technique. The change in velocity and position is based on animals' sociological instincts for hunting down their food. The DPSO is among the most important computational swarm intelligence techniques. It provides a better alternative for a complicated error surface than the gradient search-based stochastic technique by converging to the optimal global solution. Moreover, it is heuristic in nature, robust, easier to implement, effective, and reliable. The optimization function estimates the fitness value of each particle in the search space [28].

In the DPSO algorithm, any problem's probable result is a particle, considered a point in n-dimensional search space. The objective function assigns a fitness value to the particle. When searching for solutions in space after the best particle exists, each particle has a speed that affects its direction and distance. The DPSO identifies the particles that can't pick their individual best in several succeeding iterations. The stagnant particles are identified this way, and these particles require additional thrust to move them towards the optimal solution again. DPSO provides push to the stagnant particles in the direction of unexplored potential regions, which also adds diversity. It requires an external push that sends trapped particles outside the local optima position while mitigating the repercussions. The DPSO mimics the PSO's performance for all other situations and does not disturb its fast convergence property [29-36].

The problems of stagnation and local optima are removed in the dynamic PSO (DPSO) variant. It maintains the quick convergence rate by monitoring the particle's personal best positions (pBest). The algorithm monitors the number of iterations the pBest and gBest have not altered. If there is no change in the pBest and gBest for the defined iteration, they were changed back to their previous best positions. Then, the pBest and gBest are yet again tracked to observe whether they can get the targets better or not. The value replacement is made permanent if they can improve the targets or restore the old values. This process continues until the algorithm reaches the optimal point or the termination condition. Hence, it can be concluded that the Dynamic-PSO (DPSO) algorithm has primarily three significant milestones, i.e., (i) the possible unexplored potential regions are identified and preserved, (ii) The pBest and gBest of the stagnated particles are restructured, and (iii) the changes made in the pBest and gBest values are monitored and accepted.

Wind Power Forecasting Using the Proposed Hybrid Technique

1. In the first case, the MF parameters of ANFIS are optimized using DPSO. The signal coefficients obtained using wavelet transform are given as input to the optimized ANFIS models. Hence, there are four models of ANFIS for four different signal coefficients. The a3 coefficients of all four variables, for example, are fed into the ANFIS 1 model. In the same way, the other three ANFIS models will receive the coefficients d3, d2, and d1 as inputs. Figure 5 depicts the block diagram of the proposed WT-ANFIS-DPSO hybrid model.
2. The main steps for WT-ANFIS-DPSO modeling are given as follows:
3. All the variables are transformed using wavelets as it accurately observes the random variations of variables.
4. For training and testing of ANFIS, 250 samples and 24 samples of each variable are taken. The output after the wavelet transform comprises a3, d3, d2, and d1 coefficients for each variable.
5. Subsequently, the decomposition level is selected by taking a threshold value to remove the coefficients with lower values, thus improving the signal's quality by removing noise.

Average hourly data of Agar in the USA is considered for the proposed work. It was transformed using Daubechies wavelet of order four and three decomposition levels, as shown in Figures 6, 7, 8, and 9. The proposed model inputs wind direction, speed, historical wind power, and air temperature. The average hourly values of all four input variables are utilised for training the ANFIS model. The proposed work uses one dimensional (1-D) wavelet from the wavelet toolbox. Two hundred fifty (250) samples of each variable are taken for the training purpose, and twenty-four (24) samples are taken for testing. The decomposed coefficients are given as input to the optimized ANFIS models. The output of all the four ANFIS models was reconstructed using inverse wavelet transform of the same order to get the final forecasted value of wind power. Figure 10 compares the forecasted value and actual sub-series using the ANFIS model.

For training, validation, and testing of the ANFIS models, days from each year's four seasons are chosen. As a result, days with unusually high wind energy generation are eliminated. The proposed hybrid ANFIS model is found to have lower prediction error than other models. The proposed hybrid ANFIS model performs better, with average MSE and MAP Error of 2.11 and 9.38, respectively. Figures 11, 12, 13, and 14 show the proposed hybrid ANFIS model's forecasting results for the summer, winter, fall, and spring days, respectively [28].

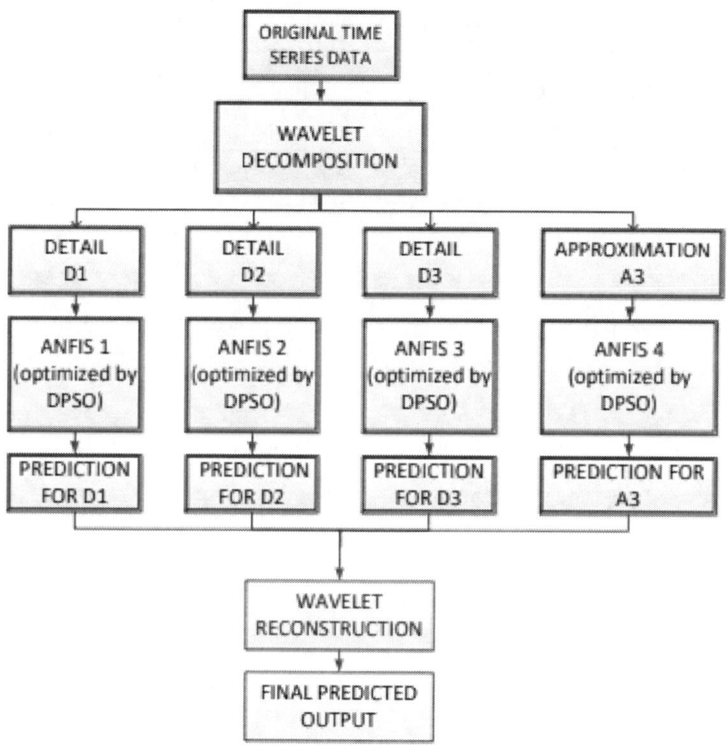

Figure 5. Proposed hybrid model.

Table 1. Comparison of the models

SEASON	WT+ANFIS+DPSO		ANFIS+DPSO		ANFIS	
	MSE	MAPE	MSE	MAPE	MSE	MAPE
Summer	1.17	9.01	2.55	9.29	4.02	12.18
Fall	2.74	9.97	3.10	11.02	4.24	12.75
Winter	2.02	9.11	2.81	9.92	3.98	12.07
Spring	2.51	9.44	2.97	10.11	3.97	12.01
Average	2.11	9.38	2.85	10.08	4.05	12.25

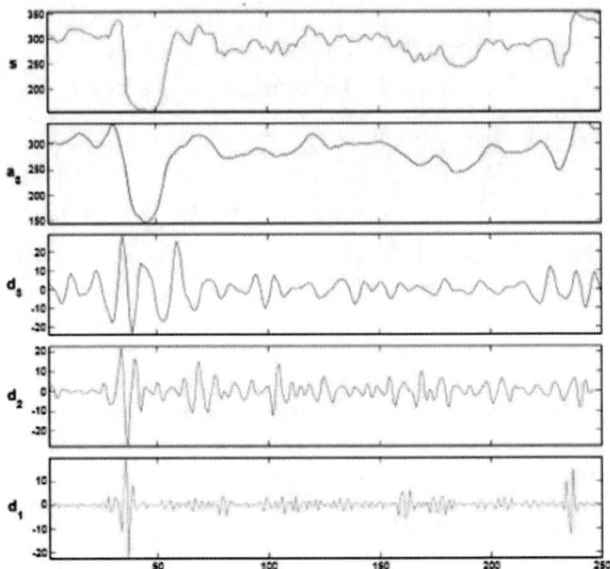

Figure 6. Three-level decomposition of wind direction variable.

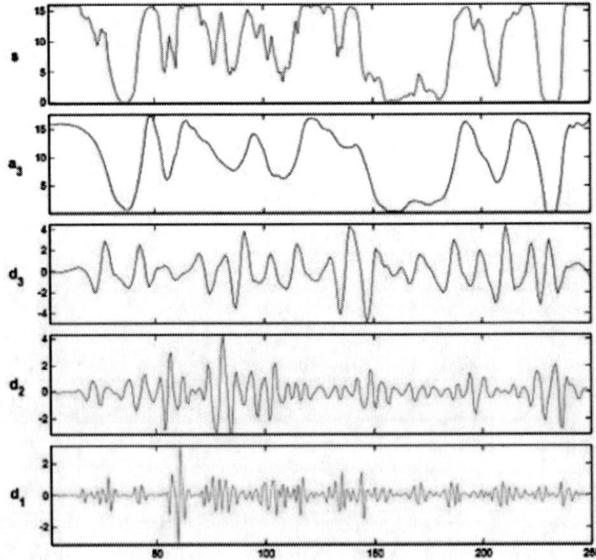

Figure 7. Three-level decomposition of wind speed variable.

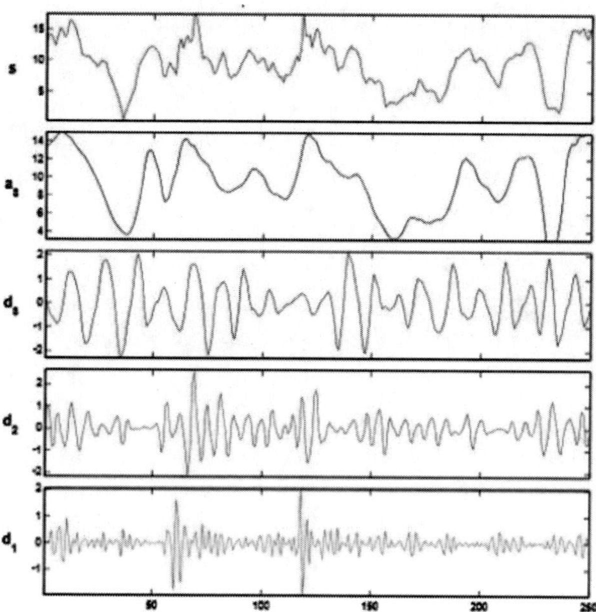

Figure 8. Three-level decomposition of wind power variable.

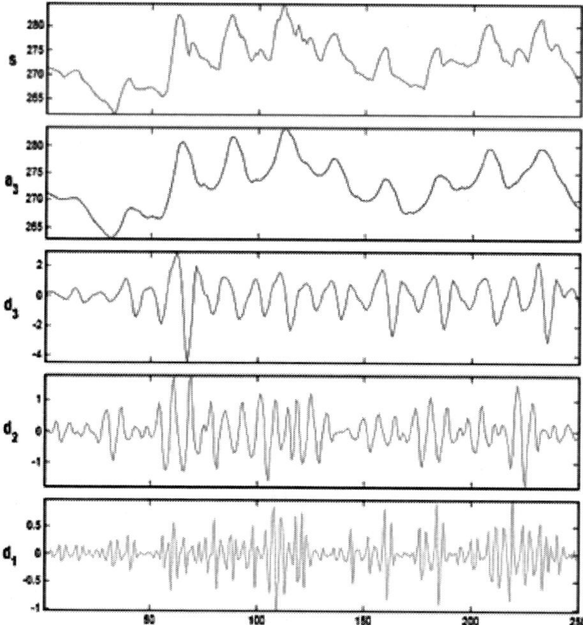

Figure 9. Three-level decomposition of the air temperature variable.

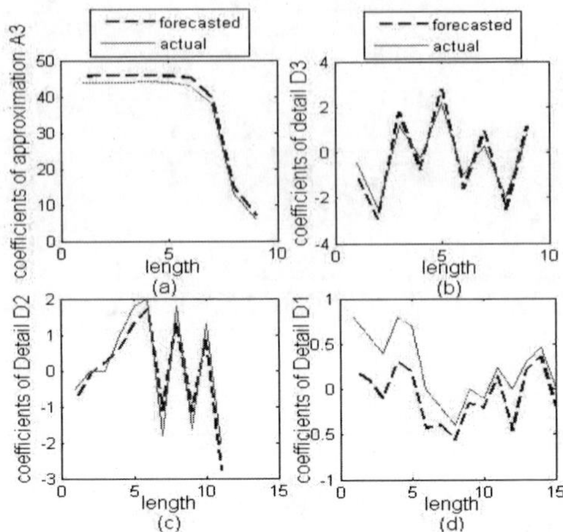

Figure 10. Actual and forecasted output of the ANFIS model.

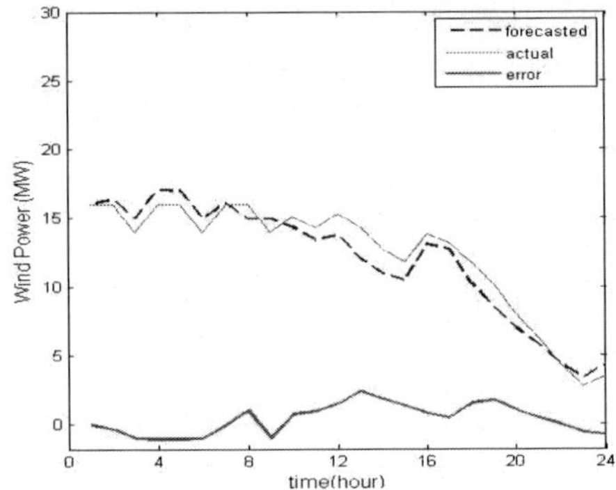

Figure 11. Actual and forecasted wind power for a summer day.

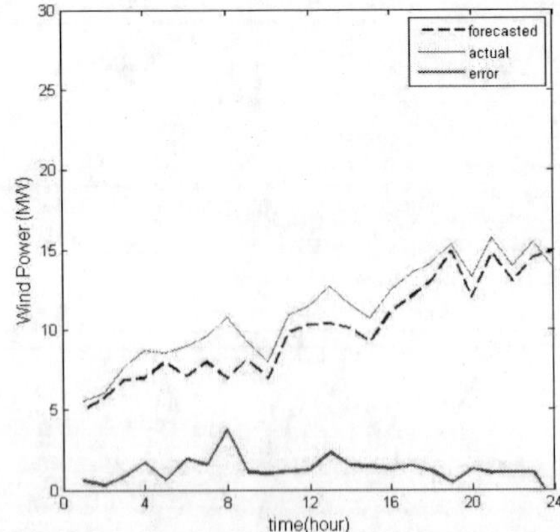

Figure 12. Wind power for a winter day (actual and forecasted).

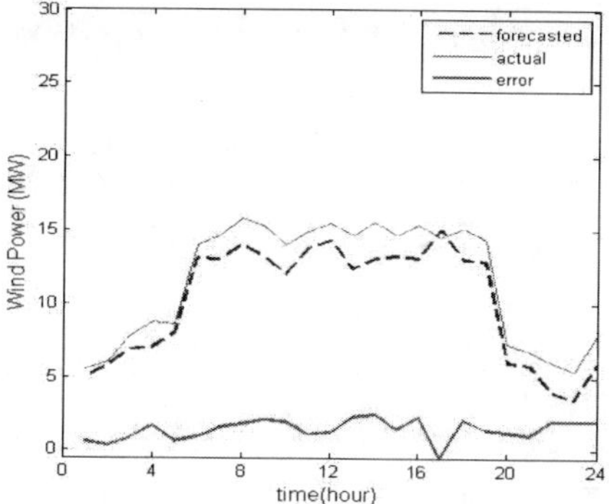

Figure 13. Wind power for a fall day (actual and forecasted).

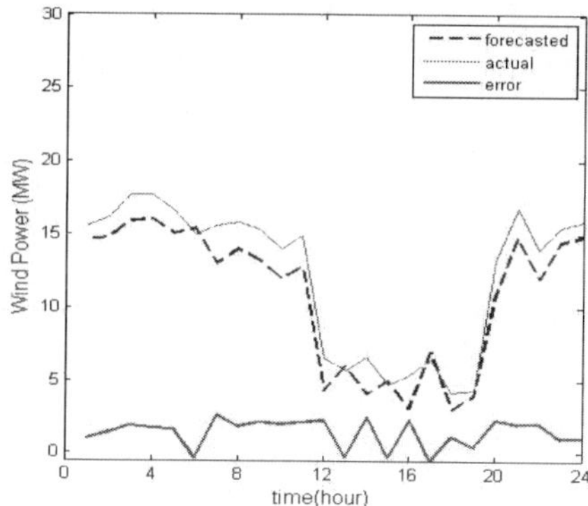

Figure 14. Wind power for a spring day (actual and forecasted).

Wind Power Prediction Using Hybrid NAR/NARX Model

In the second case, wind power (P) data is decomposed as output for NAR, and wind speed (WS) is used additionally for the NARX as exogenous input. First, both the signals were decomposed using wavelet transform. Then, each sub-series of wind power and wind speed, obtained after wavelet decomposition, are used as a target and exogenous variable for NAR and NARX models.

The network consists of input corresponding to the wind power series and the output value of prediction. After the data pre-processing and analysis stage, the value of the delay is determined using the trial and error method.

The following are the stages for modeling the proposed hybrid NAR and NARX neural network models for wind power and speed forecasting.

1. The db4 mother wavelet with two decomposition levels is used to divide the wind power and wind speed series into a set of stationary sub-series.
2. The wind power and wind speed data are normalized before training the neural networks.
3. NAR and NARX networks are formed with appropriate training for each sub-series after decomposition. Then, the models' input lags (input and feedback delays) are chosen using the ACF and PACF plots.
4. The NAR and NARX networks are employed to predict the related decomposed sub-series.
5. After obtaining predicted values for each decomposed sub-series with NAR and NARX models, they are aggregated (reconstructed) to get the final predicted value of wind power (target).

Figure 15 depicts the block diagram of the proposed hybrid NAR/NARX Model for forecasting. The activation functions for the hidden and output layers of the proposed hybrid NAR and NARX models are the 'hyperbolic' and 'linear' functions. Weights and biases are computed to optimize the performance in the training process.

Levenberg Marquardt Back Propagation (LMBP) algorithm is used for training both univariate and multivariate models. The trial and error method has been applied to select optimal input and feedback delay values and the appropriate number of hidden layers with the least mean square error value. All the training examples are passed through the learning algorithm simultaneously before modifying weights in the training process. The weights are updated to reduce the mean square error after each epoch. Multi-step ahead prediction is achieved by changing the network into a closed-loop configuration. In the training process, a set of inputs and matching outputs of historical data are provided to the network. The success of training is largely determined by the number of inputs given to the neural network. A neural network maps input and output relations to decrease the learning process by altering weights and biases the difference between generated and expected output at each step. The iterations continue until the results approach the global minima [32].

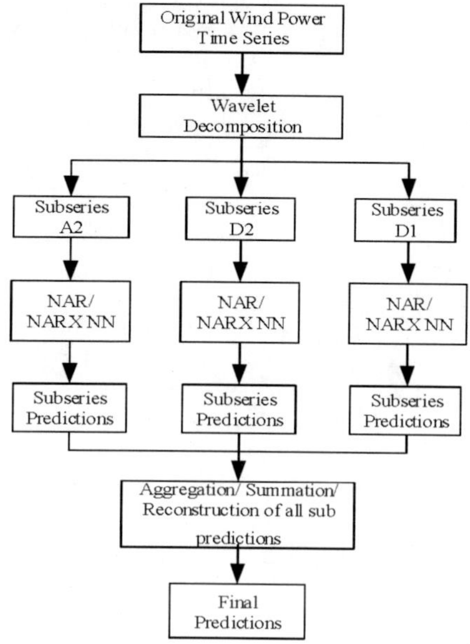

Figure 15. Block diagram of the proposed WT-NAR/NARX model.

In this case, the averaged hourly wind power and wind speed data have been taken as input. The Daubechies' db4' has been used as a mother wavelet with two decomposition levels. The A2 approximation sub-series of each time series better explained the original and resulted in the smallest error. Because high-frequency outliers are eliminated at this level, a smoother and simpler signal is achieved for forecasting. All the approximations and details sub-series were trained with NAR and NARX neural network models.

Each of the three NAR models (NAR1, NAR2, and NAR3) utilized 70%, 15%, and 15% data of decomposed sub-series of wind power time series (cA2_P, cD2_P, and cD1_P) for training, validation, and testing, respectively. Similarly, each of the three NARX models (NARX1, NARX2, and NARX3) utilized 70%, 15%, and 15% data of decomposed sub-series of wind power and wind speed (as exogenous input) time series for training, validation, and testing respectively. In three NAR network modeling, the number of feedback delays and hidden layer neurons has been selected as 4 and 12, 3 and 10, 2 and 6, respectively, using the trial and error method. Ten simulation runs for each model with varying feedback delays from 1 to 6, and hidden layer neurons ranging from 5 to 20 are carried out. The curves of training performance and regressions plots of all the three NAR models for each sub-series are shown in Figures 16, 17, and 18.

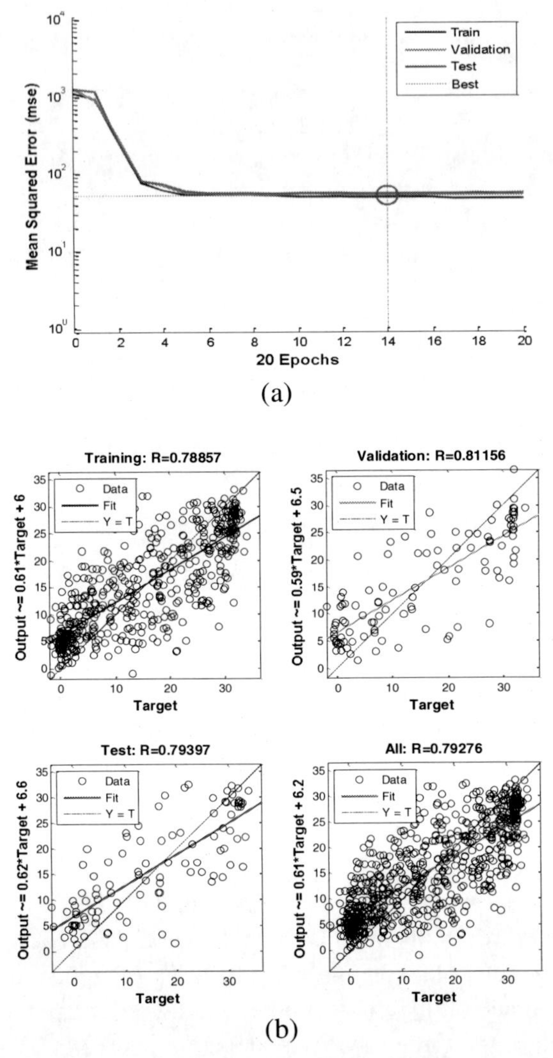

Figure 16. (a) Training performance (b) regression plot of the NAR1 model.

In NARX models (NARX1, NARX2, and NARX3), the decomposed sub-series (cA2_WS, cD2_WS, and cD1_WS) of wind speed data is taken as exogenous input to the neural networks. Due to the higher cross-correlation of wind power and wind speed, wind speed has been considered an exogenous input compared to wind direction and temperature. Along with wind speed as exogenous input, NARX uses wind power output feedback as an input. Using the trial and error method, all three models, input and feedback delays, and hidden layer neurons are selected as 7 and 12, 2 and 10, 2 and 10, respectively. For each model, ten simulation runs are performed with feedback delays ranging from 1 to 7 and hidden layer neurons ranging from 5 to 20. The training performance curves and regressions plots of all three NARX models for each subseries are shown in Figures 19, 20, and 21.

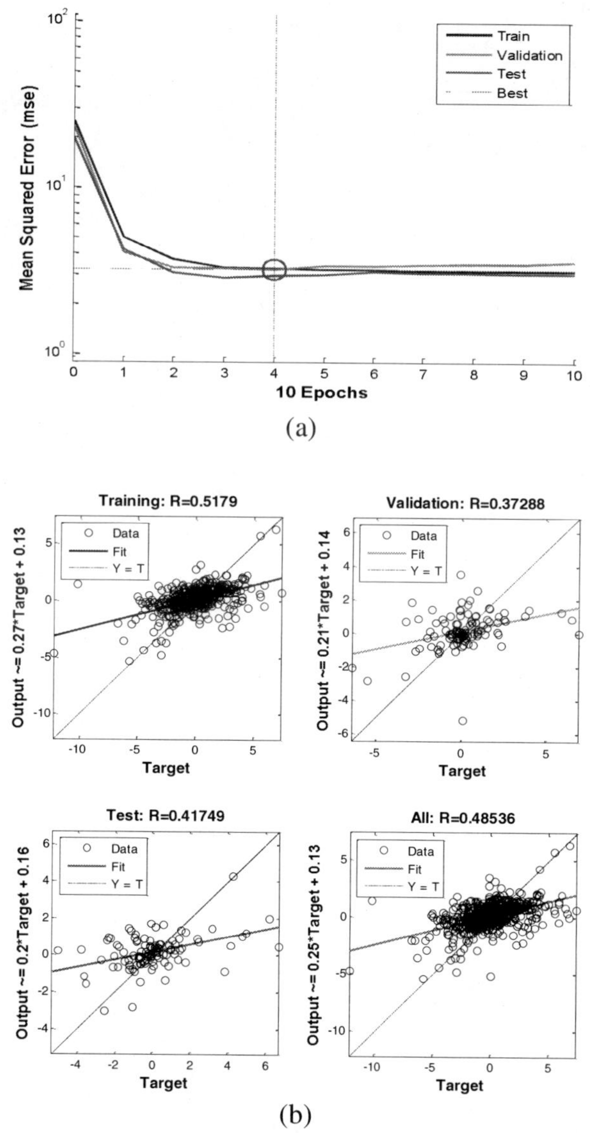

Figure 17. (a) Training performance (b) regression plot of the NAR2 model.

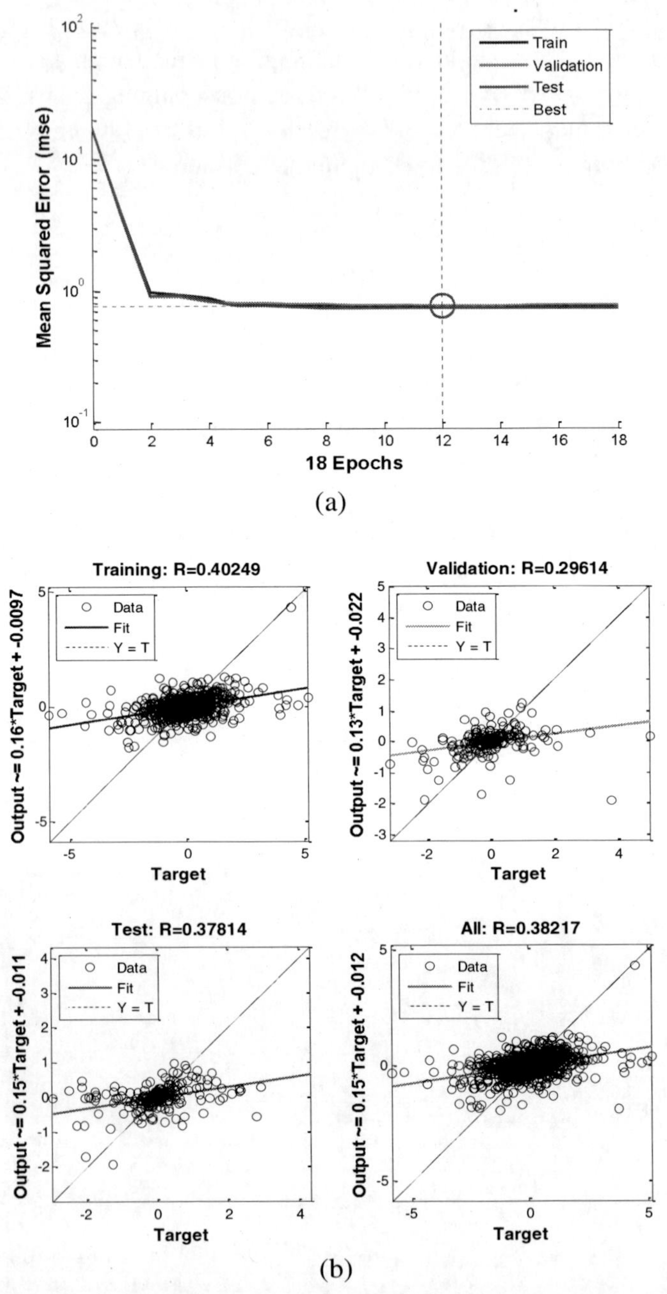

Figure 18. (a) Training performance (b) regression of the NAR3 model.

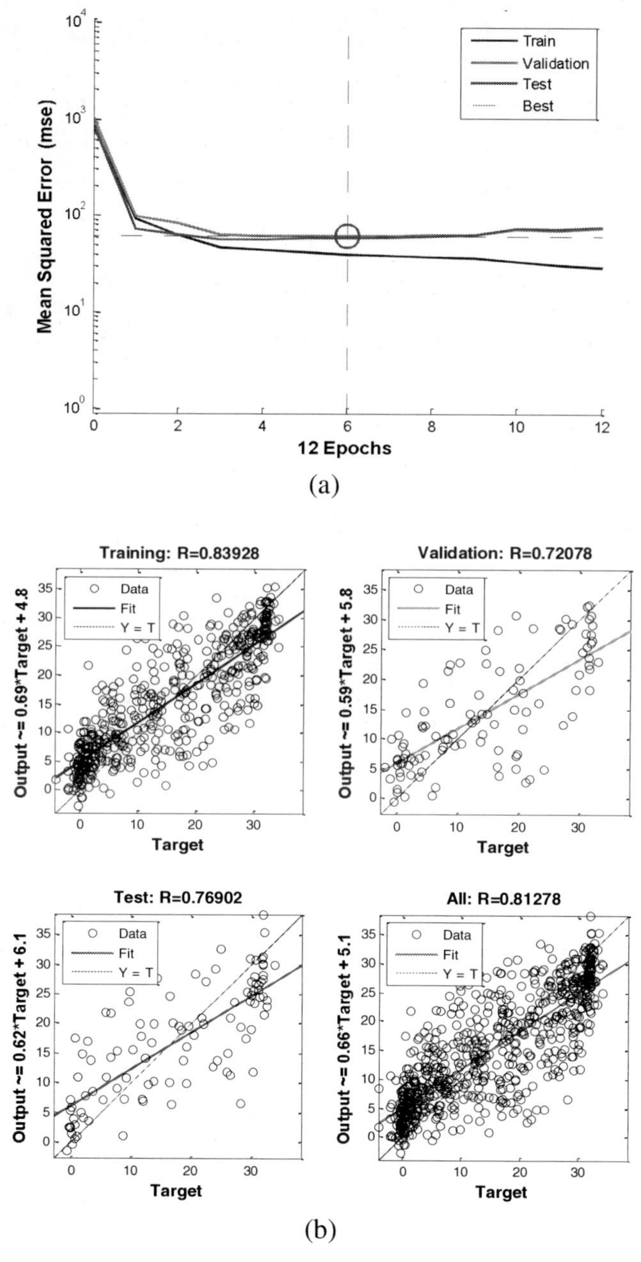

Figure 19. (a) Training performance (b) regression plot for the NARX1 model.

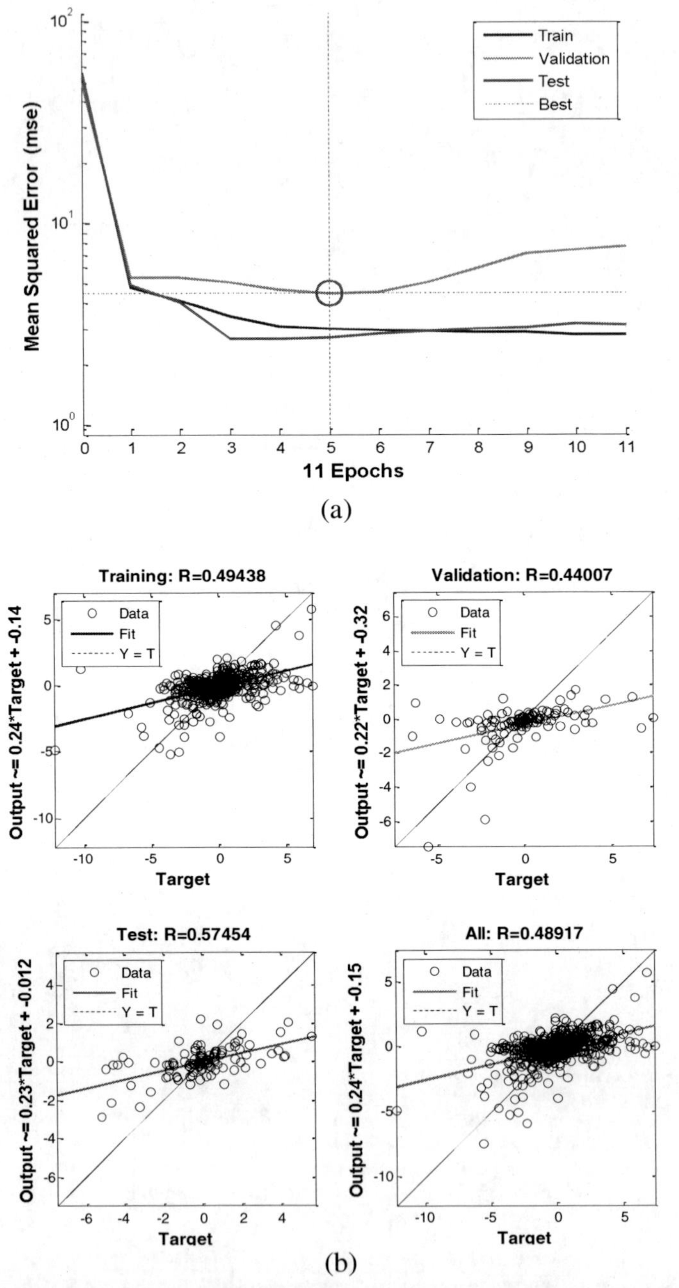

Figure 20. (a) Training performance (b) regression plot for the NARX2 model.

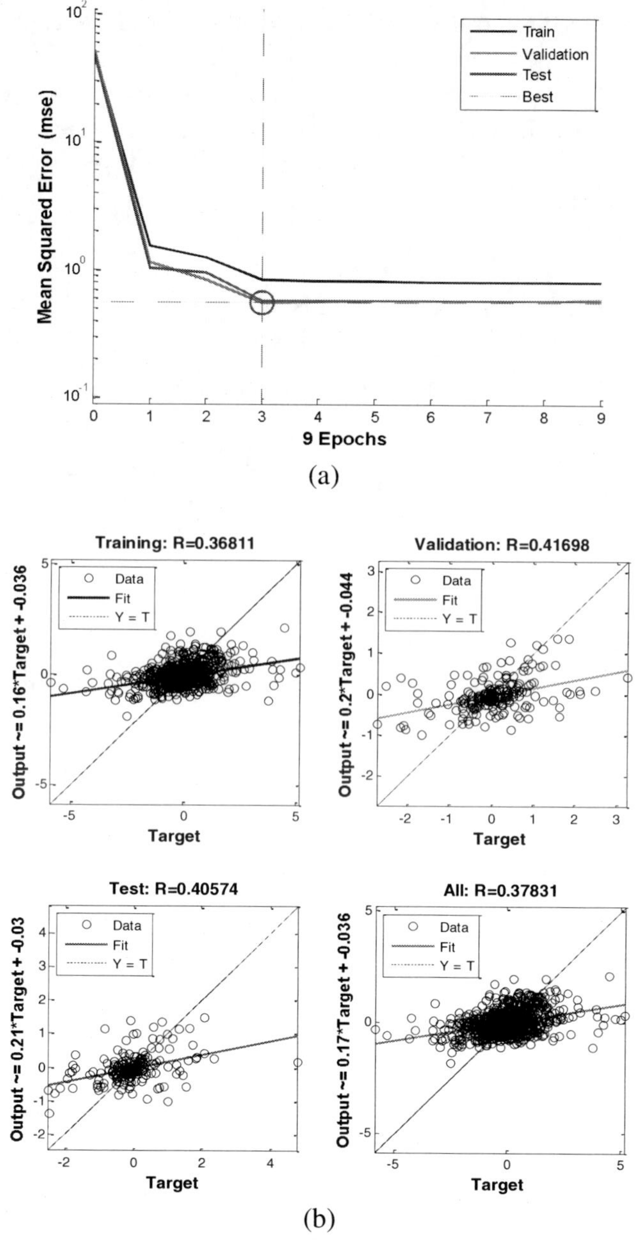

Figure 21. (a) Training performance (b) regression plot for the NARX3 model.

The day ahead forecast of each sub-series, with NAR and NARX models, has been aggregated separately to achieve the final forecast value. The proposed model's performance is compared to that of the persistence model in terms of MSE, MAE, and MAPE. Table 2 and Figure 22 show the actual and predicted values for the day ahead using the persistence, WT-NAR, and WT-NARX models. The result reveals the better performance of the WT-NARX model compared to other models.

Table 2. Performances comparison of persistence, WT-NAR, and WT-NARX models

Model	6 hours ahead			12 hours ahead			24 hours ahead		
	MAE	MSE	MAPE	MAE	MSE	MAPE	MAE	MSE	MAPE
Persistence	1.3082	3.0469	32.5466	0.8	1.5726	40.3402	1.1313	3.2788	49.5338
WT-NAR	1.484	3.327	36.1983	1.0824	2.0133	49.3279	1.1239	1.8775	40.1058
WT-NARX	0.6182	0.7123	22.6199	0.7898	0.8712	42.9186	0.8915	1.0212	37.298

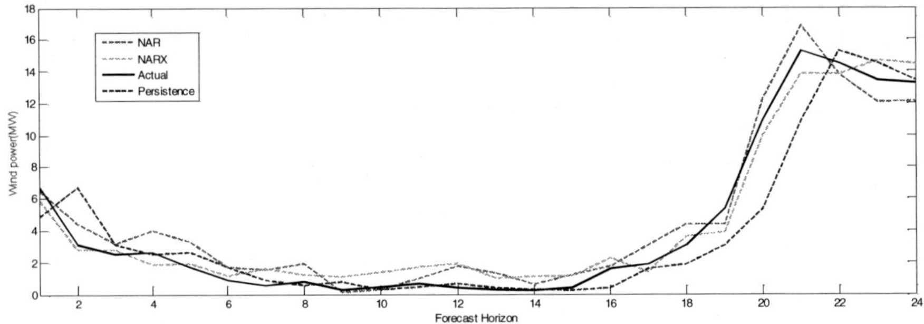

Figure 22. Actual and forecasted curves of persistence, WT-NAR, and WT-NARX models.

Conclusion

In case 1, the proposed hybrid model combines WT and an ANFIS to forecast wind power. The proposed model's forecasting accuracy is increased by optimizing its parameters using the DPSO algorithm. The DPSO algorithm also prevents the model from getting trapped into local minima. In case 2, the NAR and NARX recurrent dynamic models have been used for the short-term prediction of wavelet decomposed wind power subseries. The decomposed sub-series is fed into neural networks as input. In the NARX model, the wind speed is considered exogenous input due to its higher cross-correlation with wind power than wind direction and temperature. The simulation results of both cases show that the hybrid models possess better forecasting accuracy and are more suitable for forecasting nonlinear variables. The NARX with the exogenous input model outperformed all the other models for day-ahead prediction in the presented work. The results validate the advantage and usefulness of the proposed approaches for wind power forecasting.

References

[1] Sharma, Rahul, and Diksha Singh. "A review of wind power and wind speed forecasting." *Journal of Engineering Research and Application* 8, no. 7 (2018): 1-9.

[2] Kandil, Nahi, Rene Wamkeue, MaaroufSaad, and Semaan Georges. "An efficient approach for short-term load forecasting using artificial neural networks." *International Journal of Electrical Power & Energy Systems* 28, no. 8 (2006): 525-530.

[3] Saber, Ahmed Yousef, and Abdulaziz Al-Shareef. "Load forecasting of a desert: A computational intelligence approach." In *2009 15th International Conference on Intelligent System Applications to Power Systems*, pp. 1-7. IEEE, 2009.

[4] Ferrero Bermejo, Jesús, Juan F. Gomez Fernandez, Fernando Olivencia Polo, and Adolfo CrespoMárquez. "A review of the use of artificial neural network models for energy and reliability prediction. A study of the solar PV, hydraulic and wind energy sources." *Applied Sciences* 9, no. 9 (2019): 1844.

[5] Singh, Pavan Kumar, Nitin Singh, and Richa Negi. "Short-term wind power prediction using hybrid autoregressive integrated moving average model and dynamic particle swarm optimisation." *International Journal of Cognitive Informatics and Natural Intelligence (IJCINI)* 15, no. 2 (2021): 111-138.

[6] Giles, C. Lee, Gary M. Kuhn, and Ronald J. Williams. "Dynamic recurrent neural networks: Theory and applications." *IEEE Transactions on Neural Networks* 5, no. 2 (1994): 153-156.

[7] Yao, S. J., Y. H. Song, L. Z. Zhang, and X. Y. Cheng. "Wavelet transform and neural networks for short-term electrical load forecasting." *Energy conversion and management* 41, no. 18 (2000): 1975-1988.

[8] Soltani, Skander. "On the use of the wavelet decomposition for time series prediction." *Neurocomputing* 48, no. 1-4 (2002): 267-277.

[9] Al Wadia, M. T. I. S., and M. Tahir Ismail. "Selecting wavelet transforms model in forecasting financial time series data based on ARIMA model." *Applied Mathematical Sciences 5*, no. 7 (2011): 315-326.

[10] Mallat, Stephane G. "A theory for multi-resolution signal decomposition: the wavelet representation." *IEEE transactions on pattern analysis and machine intelligence* 11, no. 7 (1989): 674-693.

[11] Reis, AJ Rocha, and AP Alves Da Silva. "Feature extraction via multiresolution analysis for short-term load forecasting." *IEEE Transactions on power systems* 20, no. 1 (2005): 189-198.

[12] Abhinav, Rishabh, Naran M. Pindoriya, Jianzhong Wu, and Chao Long. "Short-term wind power forecasting using wavelet-based neural network." *Energy Procedia* 142 (2017): 455-460.

[13] Aggarwal, Sanjeev Kumar, Lalit Mohan Saini, and Ashwani Kumar. "Electricity price forecasting in Ontario electricity market using wavelet transform in artificial neural network based model." *International Journal of Control, Automation, and Systems* 6, no. 5 (2008): 639-650.

[14] Amjady, Nima, and Farshid Keynia. "Short-term load forecasting of power systems by combination of wavelet transform and neuro-evolutionary algorithm." *Energy* 34, no. 1 (2009): 46-57.

[15] Jang, J. S. R. "ANFIS: adaptive-network-based fuzzy inference system." *IEEE transactions on systems, man, and cybernetics* 23, no. 3 (1993): 665-685.

[16] Yun, Zhang, Zhou Quan, Sun Caixin, Lei Shaolan, Liu Yuming, and Song Yang. "RBF neural network and ANFIS-based short-term load forecasting approach in real-time price environment." *IEEE Transactions on power systems* 23, no. 3 (2008): 853-858.

[17] Karaboga, Dervis, and Ebubekir Kaya. "Adaptive network based fuzzy inference system (ANFIS) training approaches: a comprehensive survey." *Artificial Intelligence Review* 52, no. 4 (2019): 2263-2293.

[18] Shoorehdeli, Mahdi Aliyari, Mohammad Teshnehlab, Ali Khaki Sedigh, and M. AhmadiehKhanesar. "Identification using ANFIS with intelligent hybrid stable learning algorithm approaches and stability analysis of training methods." *Applied Soft Computing* 9, no. 2 (2009): 833-850.

[19] Liu, Hui, Hong-qi Tian, Chao Chen, and Yan-fei Li. "An experimental investigation of two Wavelet-MLP hybrid frameworks for wind speed prediction using GA and PSO optimisation." *International Journal of Electrical Power & Energy Systems* 52 (2013): 161-173.

[20] Maier, Holger R., and Graeme Clyde Dandy. "Determining inputs for neural network models of multivariate time series." *Computer-Aided Civil and Infrastructure Engineering* 12, no. 5 (1997): 353-368.

[21] Di Nunno, Fabio, Francesco Granata, Rudy Gargano, and Giovanni de Marinis. "Prediction of spring flows using nonlinear autoregressive exogenous (NARX) neural network models." *Environmental Monitoring and Assessment* 193, no. 6 (2021): 1-17.

[22] Prema, V., K. Uma Rao, B. S. Jnaneswar, ColathurArvindBadarish, PatilShreenidhi Ashok, and SiddarthAgarwal. "Application of hybrid neuro-wavelet models for effective prediction of wind speed." In *Intelligent Systems Technologies and Applications*, pp. 345-354. Springer, Cham, 2016.

[23] Ruiz, Luis Gonzaga Baca, Manuel PegalajarCuéllar, Miguel Delgado Calvo-Flores, and María Del Carmen Pegalajar Jiménez. "An application of non-linear autoregressive neural networks to predict energy consumption in public buildings." *Energies* 9, no. 9 (2016): 684.

[24] Pereira, Fabio Henrique, Francisco Elânio Bezerra, Shigueru Junior, Josemir Santos, Ivan Chabu, Gilberto Francisco Martha de Souza, FábioMicerino, and Silvio IkuyoNabeta. "Nonlinear autoregressive neural network models for prediction of transformer oil-dissolved gas concentrations." *Energies* 11, no. 7 (2018): 1691.

[25] Shekhawat, Anirudh S. "Wind power forecasting using artificial neural networks." *Paper id: IJERTV3IS041239, School of electrical engineering Vellore Institute of technology Chennai, India* (2014).

[26] Cadenas, Erasmo, Wilfrido Rivera, Rafael Campos-Amezcua, and Roberto Cadenas. "Wind speed forecasting using the NARX model, case: La Mata, Oaxaca, México." *Neural Computing and Applications* 27, no. 8 (2016): 2417-2428.

[27] Cadenas, Erasmo, Wilfrido Rivera, Rafael Campos-Amezcua, and Christopher Heard. "Wind speed prediction using a univariate ARIMA model and a multivariate NARX model." *Energies* 9, no. 2 (2016): 109.

[28] Parsopoulos, Konstantinos E., and Michael N. Vrahatis, eds. *Particle swarm optimisation and intelligence: advances and applications: advances and applications*. IGI global, 2010.

[29] Zhao, Hongbo, and LinaFeng. "An improved adaptive dynamic particle swarm optimisation algorithm." *Journal of Networks* 9, no. 2 (2014): 488.

[30] Garg, Harish. "A hybrid GA-GSA algorithm for optimising the performance of an industrial system by utilising uncertain data." In *Handbook of research on artificial intelligence techniques and algorithms*, pp. 620-654. IGI Global, 2015.

[31] Garg, Harish. "A hybrid PSO-GA algorithm for constrained optimisation problems." *Applied Mathematics and Computation* 274 (2016): 292-305.

[32] Zhang, Xiaoyu, Rui Wang, Tianjun Liao, Tao Zhang, and YabinZha. "Short-term forecasting of wind power generation based on the similar day and Elman neural network." In *2015 IEEE Symposium Series on Computational Intelligence*, pp. 647-650. IEEE, 2015.

[33] Eberhart, Russell, and James Kennedy. "A new optimiser using particle swarm theory." In *MHS'95. Proceedings of the sixth international symposium on micro machine and human science*, pp. 39-43. Ieee, 1995.

[34] Liang, Jing J., A. Kai Qin, Ponnuthurai N. Suganthan, and S. Baskar. "Comprehensive learning particle swarm optimiser for global optimisation of multimodal functions." *IEEE transactions on evolutionary computation* 10, no. 3 (2006): 281-295.

[35] Saxena, Nitin, AshishTripathi, K. K. Mishra, and Arun Kumar Misra. "Dynamic-PSO: An improved particle swarm optimiser." In *2015 IEEE Congress on Evolutionary Computation (CEC)*, pp. 212-219. IEEE, 2015.

[36] Chang, Wen-Yeau, Po-Chuan Chang, and Ho-Chian Miao. "Short Term Wind Power Generation Forecasting Using Adaptive Network-Based Fuzzy Inference System." In *2015 International Conference on Computational Intelligence and Communication Networks (CICN)*, pp. 1299-1302. IEEE, 2015.

Chapter 10

Design Optimization of Inner Rotor Permanent Magnet Synchronous Machine Used in Wind Energy Conversion System Using Swarm Intelligence

Vinod Puri[1,*] and Yogesh Kumar Chauhan[2]

[1]Department of Electrical Engineering, Baba Ghulam Shah Badshah University, Rajouri, Jammu and Kashmir, India

[2]Deparment of Electrical Engineering, Kamla Nehru Institute of Technology Sulatnpur, Uttar Pradesh, India

Abstract

Wind energy is one of the possible solutions by which the global warming issue is reduced. An efficient system is required to harness the wind energy using Wind Energy Conversion System (WECS). The generator is the main part of WECS. It acts as a conversion device which converts the wind energy into the electrical energy. Therefore, there is a need to design such generators which are efficient and provide an affordable solution to the current problem. Nowadays Permanent Magnet Synchronous Generators (PMSM) are immensely used in the WECS. The configuration of PMSM used in this chapter is Inner Rotor Permanent Magnet Synchronous Machine (IRPMSM). The chapter puts emphasis on the design optimization of IRPMSM for WECS. A 500 kVA, 50Hz, 3.3kV, 600 rpm IRPMSM has been designed and optimized using artificial intelligence techniques. The swarm intelligence techniques used to optimize the IRPMSM are Gravitational Search Algorithm (GSA) and its hybridization with Particle Swarm Optimization (PSO). This process

*Corresponding Author's Email: vinodpuri@bgsbu.ac.in.

In: Applied Artificial Intelligence (AI) to Green Power Technology
Editors: Yogesh Kumar Chauhan, Ranjan Kumar Behera and Asheesh K. Singh
ISBN: 979-8-88697-131-6
© 2022 Nova Science Publishers, Inc.

can be used as a first step giving a close estimate regarding the material required, cost and performances of the machine.

Keywords: optimization, PMSM, machine design methodology, GSA, GSA-PSO, WECS

Introduction

Many researchers and scientists are searching for a possible solution to reduce the greenhouse gases in the environment and lowering down the global warming effect. World is moving towards the renewable sources of energy and have a check on conventional sources of energy. The wind energy will provide the necessary solution to mitigate the problem. Many new technologies are used to enhance the efficiency of WECS,(Grabic, Celanovic, & Katic, 2008; Lahda, Mouradi, El Hibaoui, & Mimet, 2013). Generally, induction generators are used in the WECS but nowadays PMSGs are immensely used to enhance the efficiency of WECS.

Due to permanent magnets, there is flexibility in the design of the machine (Boldea, Tutelea, & Pitic, 2004; Fitan, Messine, & Nogarede, 2003; Hosseini, Agha-Mirsalim, & Mirzaei, 2007; Say & Say, 1995). Many new designs have been developed like axial flux permanent magnet synchronous machine (AFPMSM) (Chan, Wang, & Lai, 2010; Hosseini et al., 2007), radial flux permanent magnet synchronous machine (RFPMSM), Permanent magnet linear generator (PMLG) (Hsiao, Lai, Zheng, & Li, 2021). As there are different designs the same is the case with applications. It is mainly used in power generation, automation, electrical vehicles (Aydin & Guven, 2013), space application (Yuan et al., 2021) etc.

Depending up on the applications, there are modifications in the design of PMSM which can be analyzed in the predesign studies, where the estimation of material required, performance of the machine have been evaluated and optimal design parameters have been estimated while satisfying all the design constraints (Fitan et al., 2003; Gope & Goel, 2021). These studies can be done while modeling the machine into a mathematical problem and solving it by using optimization techniques. Some simulation software are also used to validate the design problem and to carry FEM analysis etc. (Kant, Kirtley, Iyer, & Schlager, 2021).

There are several optimization techniques which are used to optimize the design problem as reported in the literature like Genetic Algorithm (Mohd-

Shafri, Tiang, Tan, Ishak, & Ahmad, 2022; Roomi, Vahedi, & Mirnikjoo, 2021), Contour Algorithm (Wi & Lim), Neural Network (Yuan et al., 2021), Tuguchi Method (Gope & Goel, 2021), Finite Element Simulation Based Optimization (Kant et al., 2021), Grey Wolf Optimization Algorithm, Culture Algorithm (Xing & Gao, 2014), Corona Virus Heard Immunity Optimization (Kumar, Magdalin Mary, & Gunasekar, 2022), Booth's Algorithm (Asef, Perpina, Barzegaran, Lapthorn, & Mewes, 2017)etc..

In this chapter, design studies have been performed on a PMSM having specifications as 500 kVA, 3.3kV, 50 Hz, 600 rpm. The objective function has been formulated with the constraints and optimized using Gravitational Search Algorithm and its hybridization with Particle Swarm Optimization. The GSA is based on the Newton's law of gravitational forces and was proposed by Rashedi in 2009 (Rashedi, Nezamabadi-Pour, & Saryazdi, 2009). In this optimization technique the solution is having different agents called as masses. The obtained solution is a group of masses which exert the gravitational forces on each other. The solution having largest mass represents the optimal solution in the search space and it cannot be attracted by other small masses (Duman, Sönmez, Güvenç, & Yörükeren, 2012; Rashedi, Nezamabadi-Pour, & Saryazdi, 2011; Xing & Gao, 2014).

The chapter has been organized as follows. Firstly, the design problem has been formulated and secondly, the optimization algorithms have been discussed. Further in the end the optimization technique is applied to the formulated objective function and the results have been discussed in detail.

Problem Formulation

Design Problem

The objective function consist of the main function which in case of IRPMSM is the weight of IRPMSM and is formulated as

$$F(t) = W_T + \sum_{i=1}^{N} \lambda_i g(i) + \sum_{j=1}^{m} \lambda_j h(j), \tag{1}$$

$$W_s = 0.9\rho_i L\big(0.25\pi(D_o + D)(D_o - D) - (S_s w_s h_s)\big), \tag{2}$$

$$W_r = 0.225\pi\rho_i L(D - 2l_g)^2 - 0.9\rho_i \pi L\big(w_p h_p\big), \tag{3}$$

$$W_c = 2\rho_c P m_s A_c T_{ph} l_{mt}. \tag{4}$$

where, $W_T = W_s + W_r + W_{c.}$ and (5)

Where W_s is weight of stator, W_r is weight of rotor, W_c is weight of copper, W_T is the total weight of the machine, λ_i and λ_j are the Lagrange's multiplier, *g(i)* is the equality constraints, *h(i)* is the inequality constraints D_o is the outer diameter, *D* is the air-gap diameter, *L* is length of the machines, S_s is the stator slots, w_s is the width of the slot, h_s is the depth of the slot, l_g is the air gap length, w_p width of the pole, h_p height of the pole, *P* no. of poles, T_{ph} is the turns per pole per phase, l_{mt} is the length of mean turn, m_s is the no of phases, ρ_i is the density of iron, ρ_c is the density of copper.

The inequality constraints used in the formulation of the design of IRPMSM have been given below:

Efficiency (η_i)

$$\eta_i^{min} \leq \eta_i \leq \eta_i^{max}, \tag{6}$$

$$\eta_i = \frac{P_o}{(P_o + Iron\ loss + Copper\ loss + Windage\ and\ friction\ loss)}. \tag{7}$$

Here, P_o is the output power, η_i^{min} is the limit of minimum efficiency and η_i^{max} is the limit of maximum efficiency.

- Regulation (Reg_i)

$$Reg_i^{min} \leq Reg_i \leq Reg_i^{max}, \tag{8}$$

$$Reg_i = \left(\frac{E_o - E_{ph}}{E_{ph}}\right) \times 100, \tag{9}$$

Here, Reg_i^{min} is the limit of minimum regulation and Reg_i^{max} is the limit of maximum regulation

In the equation (9) no load voltage, E_o, has been calculated as,

$$E_o = (E_{Ph} + \left(I_{ph}(R\ cos\theta + X_\iota sin\theta)\right))\frac{B_{av}}{B_g}. \tag{10}$$

B_{av} is average flux density, B_g is maximum flux density, E_o is the no load voltage, Eph is the phase voltage, Iph is the phase current, R is the resistance, XL is the inductive reactance, θ is the power factor angle.

- Temperature rise (T_i)

$$T_i^{min} \leq T_i \leq T_i^{max}, \quad (11)$$

$$T_i = \frac{Q}{\frac{S_O}{c_O}+\frac{S_i}{c_i}+\frac{S_d}{c_d}}. \quad (12)$$

Here, T_i^{min} is the limit of minimum temperature and T_i^{max} is the limit of maximum temperature Q is the loss to be dissipated (watts), S_o outer surface area, S_i inner surface area, S_d surface area of the duct and c_o, c_i and c_d are the cooling coefficients.

- Flux density in stator tooth (B_{Ti})

$$B_{Ti}^{min} \leq B_{Ti} \leq B_{Ti}^{max}, \quad (13)$$

$$B_{Ti} = \frac{1.36P\phi_p}{k_i S w_t}. \quad (14)$$

Here, S is the number of slots, φ_p is the flux per pole, k_i is the stacking factor of the machine and w_t is the width of the machine, B_{Ti}^{min} is the limit of minimum temperature, B_{Ti}^{max} is the limit of maximum temperature. The above inequality constraints can be handled in the objective function using below mention formulation as:

- For inequality constraint:

$$\alpha = max(inequality\ constraint, -\lambda(iter-1)/(2r), \quad (15)$$

$$\lambda(iter) = \lambda(iter-1) + (2r\alpha). \quad (16)$$

Here, λ is the lambda which is used for up gradation, r is the penalty parameter. There are no equality constraints in the given design problem. If the problem includes the equality constraints, it can be handled by the given formulation as:

- For equality constraint:

$$\lambda(iter) = \lambda(iter - 1) + (2r(g(i))). \tag{17}$$

Optimizing Techniques

Algorithm of GSA and GSA-PSO Technique

The GSA and GSA-PSO have been discussed herewith in following section. The algorithm for the GSA and GSA-PSO have been given below,

- Algorithm for GSA

Step 1: Consider a system with N agents and the position of the i[th]agent is defined as,

$$P_i = (p_i^1, \dots\dots p_i^d, \dots\dots p_i^n); i = 1,2 \dots\dots n \tag{18}$$

Step 2: Formulating and evaluating the fitness value of the objective function. This objective function may be a minimization problem, i.e.,

$$Fitness_i = f(p_i^1, \dots\dots p_i^d, \dots\dots) = f(p_i) \tag{19}$$

Step 3: The gravitational mass and the inertial mass depend upon the fitness value of the objective function and are given by following equations as,

$$M_{ai} = M_{pi} = M_{ii} = M_i = Inetrial\ mass \tag{20}$$

$$m_i(t) = \frac{fitness_i(t) - worstfitness(t)}{bestfitness(t) - worstfitness(t)}, \tag{21}$$

Here, $fitness_i(t)$ acts as the fitness value of the object i at time t and *worstfitness(t)* and *bestfitness(t)* are given as,

$$M_i(t) = \frac{m_i(t)}{\sum_{j=1}^{N} m_j(t)}, \tag{22}$$

$$bestfitness(t) = \min_{j\in(1......N)} fitness_j(t), \tag{23}$$

$$worstfitness(t) = \max_{j\in(1......N)} fitness_j(t). \tag{24}$$

Step 4: Calculation of force acted on mass *i* from mass *j*. This force is given as,

$$F_{ij}^{d} = G(t) \times \frac{M_{pi}(t) \times M_{aj}(t)}{R_{ij}+\varepsilon} \times (p_j^d(t) - p_i^d(t)), \tag{25}$$

Here, ε is the small positive number; M_{aj} acts as active gravitational mass of object *j*; M_{pi} is the passive gravitational mass of object *i*. The $R_{ij}(t)$ is given by,

$$R_{ij}(t) = \| P_i(t), P_j(t) \| , \tag{26}$$

Step 5: Calculate the total force acting on $mass_i$ in the dth dimension in time *t* is given as follows,

$$F_{Ti}^{d}(t) = \sum_{j\in Kbest, j\neq i}^{N} rand_j \times F_{ij}^{d}(t), \tag{27}$$

where, $rand_j$ is a random number in the interval [0, 1], *Kbest* is the set of initial K *objects* with the best fitness value.

Step 6: Calculate the acceleration of mass *i* in time t in the dth dimension is given as follows,

$$a_i^d = \frac{F_{Ti}^{d}(t)}{M_i(t)}, \tag{28}$$

Step 7: The velocity and position of agents are updated as,

$$u_i^d(t+1) = rand_i \times u_i^d(t) + a_i^d(t), \tag{29}$$

$$p_i^d(t+1) = p_i^d(t) + u_i^d(t+1). \tag{30}$$

Here, u_i^d and p_i^d are the velocity and position up-gradation equation

- Algorithm for GSA-PSO

The idea of GSA-PSO is to combine the ability of social thinking gbest in PSO with the local search capability of GSA. The steps 1 to step 6 have been same except the up gradation of velocity and position of agents are updated as,

$$u_i^d(t+1) = wu_i^d(t) + c_1 \times rand \times a_i^d(t) + c_2 \times rand \times (Xgbest - X_i), \tag{31}$$

$$p_i^d(t+1) = p_i^d(t) + u_i^d(t+1). \tag{32}$$

Here, c1 and c2 are the social constants whose value ranges from 1 to 0.5.

Result and Discussion

Here, the weight of copper and iron are taken as objective function and is minimized using GSA and GSA-PSO. As stated earlier, IRPMSM is considered and its optimized results are obtained and compared with other hybrid technique. A 3.3kV, 500KVA, 3-phase, 50 Hz, 600 rpm, direct derived IRPMSM is designed as inner rotor configuration. The Table 1 and Table 2 shows the results of the optimized design parameters of IRPMSM and performance indices respectively like efficiency, regulation, temperature rise etc. The results of Zangwill's nonlinear programming using penalty function have also been included for the validation and comparison purpose (Rao, 2019).

The results are compared with the initial performance indices which are the results obtained using classical methods and some are the range of parameters. When the results of GSA are compared with the initial performance indices; the weight is reduced by 33.1264%; voltage regulation is improved by 2.506%; maximum flux density in stator tooth is reduced by 34.97%; efficiency is increased by 2.10% and temperature is reduced by 5.8%.

In case of GSA-PSO the weight is reduced by 33.621%; voltage regulation is improved by 2.739%; maximum flux density in stator tooth is reduced by 36.37%; efficiency is increased by 2.15% and temperature rise is reduced by 6.25%. It is seen that the GSA-PSO gives improved results than GSA for single objective optimization.

Table 1. Optimal design data for inner rotor PMSM

Design parameters Of IRPMSM	Range of Design Parameters	Optimal design Zangwill's NLP	Optimal design using GSA	Optimal design using GSA-PSO
CPU Time (sec)	-	-	1.93611	1.93315
Number of iterations	-	-	48	20
Outer diameter of stator (m)	1.142-1.04	1.076	1.0707	1.0723
Diameter of air gap (m)	0.9-0.8	0.80	0.8192	0.8029
Length of core (m)	0.45-0.35	0.35	0.3510	0.3504
Number of slots	90	90	90	90
Number of turns	150	150	150	150
Depth of stator slots (m)	0.0475-0.0411	0.0411	0.0443	0.0444
Width of stator slots (m)	0.013	0.013	0.013	0.013
Peak flux density in air gap (Wb/m^2)	0.719	0.719	0.719	0.719
Length of air gap (m)	0.006-0.005	0.005	0.0054	0.0050
Maximum air gap (m)	0.106-0.105	0.105	0.1054	0.1057
Height of pole (m)	0.1	0.1	0.1	0.1
Width of the pole (m)	0.05	0.05	0.05	0.05

When the results of GSA-PSO are compared with GSA the weight is further reduce by 0.74%; voltage regulation is reduced by 0.21%; maximum flux density is reduced by 2.15%; efficiency is increased by 0.021% and temperature is reduced by 0.46%.

It can be inferred from the above discussion that the GSA-PSO optimizes the single objective function (weight of IRPMSM) better than the GSA and gives improved results.

Table 2. Performance indices of the optimized machine

Performance Indices	Initial Performance Indices	Optimal design Zangwill's NLP	Optimal design using GSA	Optimal design using GSA-PSO
Regulation at 0.8pf	32.12	30.73	31.3148	31.2447
Maximum flux density in teeth (Wb/m^2)	1.92-2.1	1.45	1.3656	1.3362
Efficiency	93.49	92.02	95.482	95.503
Weight of copper (kg)	242	220	202.444	201.2033
Weight of stator (kg)	1431	1417	798.956	855.3967
Weight of rotor (kg)	1248	1236	951.975	882.321
Total weight (kg)	2921	2873	1953.375	1938.921
Temperature Rise (°C)	33.266	38.56	31.3304	31.1862

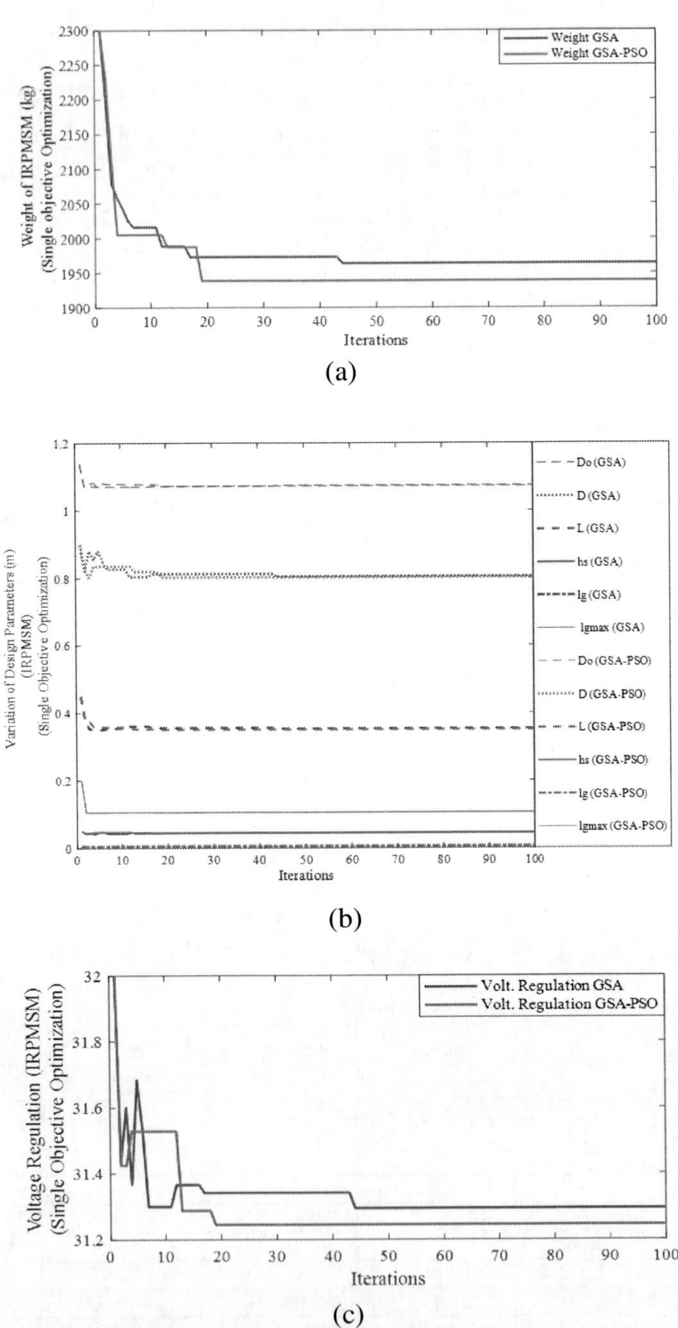

Figure 1. (Continued).

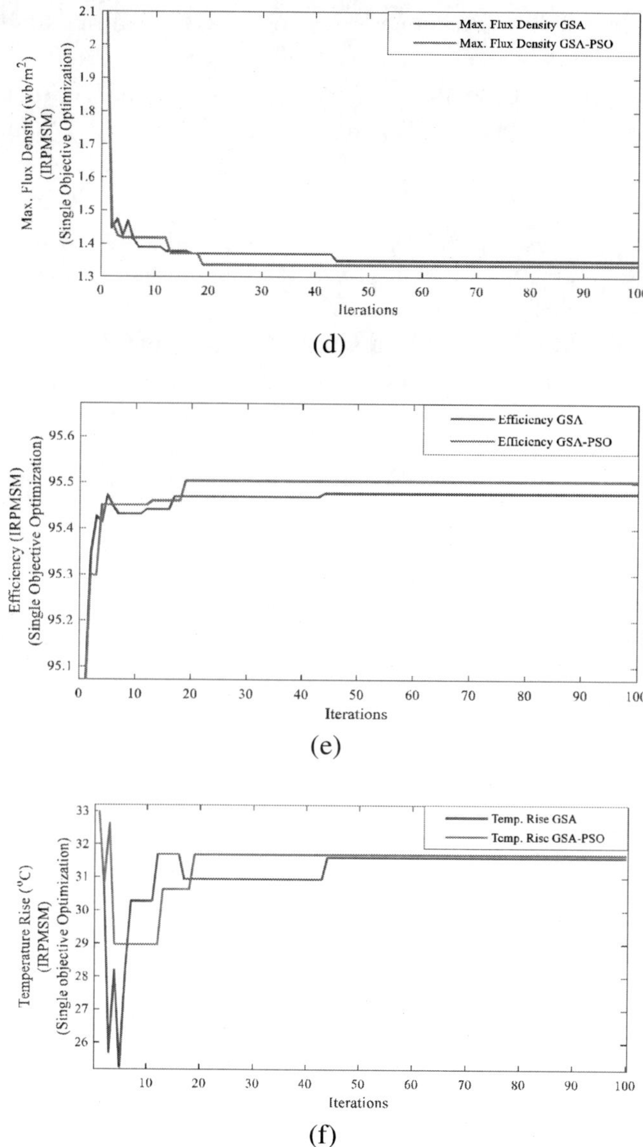

Figure 1. Performance optimized curves of single objective function of IRPMSM using GSA and GSA-PSO. (a) Comparative curve of weight of IRPMSM. (b) Variation of all the design parameters of IRPMSM. (c) Comparative curve of voltage regulation of IRPMSM. (d) Comparative curve of maximum flux density in stator tooth of IRPMSM. (e) Comparative curve of efficiency of IRPMSM (f) Comparative curve of temperature rise of IRPMSM.

The Figure 1 shows the comparative optimized results of single objective function using GSA and GSA-PSO for IRPMSM. The fig 1(a) shows the comparative optimized curve of weight function. It is seen that the curve optimize in 45 iterations for GSA and 20 iterations for GSA-PSO to converge. It shows fast convergence to optimize single objective optimization of IRPMSM.

The Fig 1 (b) shows the variation of the design variables i.e., *Do, D, L, hs, lg* and *lgmax* which vary initially to search the optimal value as the function optimize all the design parameters get settled to their optimal value. The fig 1 (c-f) shows the curves of voltage regulation, maximum flux density, efficiency and temperature rise respectively. It has been observed from the curves that the GSA-PSO gives better optimized results than GSA.

Conclusion

It is evident that the predesign process is important to find out the performance of the machine before manufacturing so that necessary modification may be done for the desired results. As the PMSM has been designed for WECS and the objective is to optimize the design for reduced weight. Following are the inferences:

- The comparative analysis has been done for single objective optimization and the results obtained by the GSA-PSO are better than the GSA.
- The weight of the machine is reduced by 33.1264% from its initial design and the whole performance of the machine has been improved.
- As compared to its initial design the wind energy needed to rotate the turbine as well as the rotor is comparable lesser.
- The above mentioned research increases the scope of optimizing algorithms namely GSA and GSA-PSO in optimizing the machine design problems and other complex problems.

References

Asef, P., Perpina, R. B., Barzegaran, M., Lapthorn, A., & Mewes, D. (2017). Multiobjective design optimization using dual-level response surface methodology and booth's algorithm for permanent magnet synchronous generators. *IEEE transactions on energy conversion, 33*(2), 652-659.

Aydin, M., & Guven, M. (2013). Design of several permanent magnet synchronous generators for high power traction applications. *Paper presented at the 2013 International Electric Machines & Drives Conference*.

Boldea, I., Tutelea, L., & Pitic, C. I. (2004). PM-assisted reluctance synchronous motor/generator (PM-RSM) for mild hybrid vehicles: electromagnetic design. *IEEE Transactions on Industry Applications, 40*(2), 492-498.

Chan, T., Wang, W., & Lai, L. (2010). Performance of an axial-flux permanent magnet synchronous generator from 3-D finite-element analysis. *IEEE transactions on energy conversion, 25*(3), 669-676.

Duman, S., Sönmez, Y., Güvenç, U., & Yörükeren, N. (2012). Optimal reactive power dispatch using a gravitational search algorithm. *IET generation, transmission & distribution, 6*(6), 563-576.

Fitan, E., Messine, F., & Nogarede, B. (2003). A general analytical model of electrical permanent magnet machine dedicated to optimal design. *COMPEL-The international journal for computation and mathematics in electrical and electronic engineering*.

Gope, D., & Goel, S. K. (2021). Design optimization of permanent magnet synchronous motor using Taguchi method and experimental validation. *International Journal of Emerging Electric Power Systems, 22*(1), 9-20.

Grabic, S., Celanovic, N., & Katic, V. A. (2008). Permanent magnet synchronous generator cascade for wind turbine application. *IEEE Transactions on Power Electronics, 23*(3), 1136-1142.

Hosseini, S. M., Agha-Mirsalim, M., & Mirzaei, M. (2007). Design, prototyping, and analysis of a low cost axial-flux coreless permanent-magnet generator. *IEEE Transactions on Magnetics, 44*(1), 75-80.

Hsiao, C.-Y., Lai, C.-H., Zheng, Z.-X., & Li, G.-Y. (2021). Design and Implement of Three-Phase Permanent-Magnet Synchronous Wave Generator using Taguchi Approach. *Energies, 14*(7), 2010.

Kant, K., Kirtley, J. L., Iyer, L. V., & Schlager, G. (2021). Finite Element Simulation-Based Design Optimization of Permanent Magnet Motors considering Drive Cycle. *SAE International Journal of Electrified Vehicles, 10*(2), 157.

Kumar, C., Magdalin Mary, D., & Gunasekar, T. (2022). MOCHIO: a novel Multi-Objective Coronavirus Herd Immunity Optimization algorithm for solving brushless direct current wheel motor design optimization problem. *Automatika, 63*(1), 149-170.

Lahda, M., Mouradi, A., El Hibaoui, A., & Mimet, A. (2013). Design and construction of a synchronous wind turbine. *Paper presented at the 2013 International Renewable and Sustainable Energy Conference (IRSEC)*.

Mohd-Shafri, S. A., Tiang, T. L., Tan, C. J., Ishak, D., & Ahmad, M. S. (2022). GA Optimization for Regression Modeling of Electromagnetic Performances Predicted by a Subdomain Model for SMPMSM in an Electric Vehicle. *Engineering Proceedings, 12*(1), 73.

Rao, S. S. (2019). *Engineering optimization: theory and practice*: John Wiley & Sons.

Rashedi, E., Nezamabadi-Pour, H., & Saryazdi, S. (2009). GSA: a gravitational search algorithm. *Information sciences, 179*(13), 2232-2248.

Rashedi, E., Nezamabadi-Pour, H., & Saryazdi, S. (2011). Filter modeling using gravitational search algorithm. *Engineering Applications of Artificial Intelligence, 24*(1), 117-122.

Roomi, F. F., Vahedi, A., & Mirnikjoo, S. (2021). Multi-Objective Optimization of Permanent Magnet Synchronous Motor Based on Sensitivity Analysis and Latin Hypercube Sampling assisted NSGAII. *Paper presented at the 2021 12th Power Electronics, Drive Systems, and Technologies Conference (PEDSTC).*

Say, M., & Say, M. (1995). *Performance and design of AC machines*: English LB S.

Wi, C.-H., & Lim, D.-K. Surrogate Assisted Contour Algorithm for Optimal Design of IPMSM. *Paper presented at the 2021 24th International Conference on Electrical Machines and Systems (ICEMS).*

Xing, B., & Gao, W.-J. (2014). *Innovative computational intelligence: a rough guide to 134 clever algorithms* (Vol. 62): Springer.

Yuan, G., Yang, T., Bozhko, S., Wheeler, P., Dragicevic, T., & Gerada, C. (2021). Neural Network aided PMSM multi-objective design and optimization for more-electric aircraft applications. *Chinese Journal of Aeronautics*. https://doi.org/10.1016/j.cja.2021.08.006.

Chapter 11

A Novel Voltage Stability Index and Application of Machine Learning Algorithm for Assessment of Voltage Stability

G. Sandhya Rani[1,*], M. Chakravarthy[1] and B. Mangu[2]

[1]Department of Electrical and Electronics Engineering,
Vasavi College of Engineering, Hyderabad, India
[2]Department of Electrical Engineering,
Osmania University, Hyderabad, India

Abstract

The present paper aims at the development of a modified voltage stability index [MVSI] for the assessment of the voltage stability state of a given power system. It also proposes the Gaussian process regression machine learning algorithm for online assessment of the voltage stability. The MVSI method and the machine learning algorithm are tested for an IEEE 30-bus test system subjected to different possible operating conditions. A comparison is made between the results of both the modified methods and the existing methods. The betterment of results with the proposed methods has been clearly highlighted.

Keywords: voltage stability, modified voltage stability index (MVSI), power system stability, voltage collapse, Gaussian process regression machine learning algorithm

---

*Corresponding Author's Email: g.sandhyarani@staff.vce.ac.in.

In: Applied Artificial Intelligence (AI) to Green Power Technology
Editors: Yogesh Kumar Chauhan, Ranjan Kumar Behera and Asheesh K. Singh
ISBN: 979-8-88697-131-6
© 2022 Nova Science Publishers, Inc.

Introduction

Blackouts take place in power systems across the world because of voltage instability. A big blackout occurred in Tokyo on 23rd July 1987, and it lasted for more than 3 hours. Many blackouts happened in UK, Sweden, Denmark, Italy and the US in 2003 [1-2].

Whenever there is any abnormality, the power system needs a continuous observation and instability prediction immediately. In the recent research studies, various methods have been applied for voltage stability and controllability. Ajjavarapu and Christy studied steady state voltage stability based on the continuous power flow method [3] and discussed a method to find continuous power flow solutions from the base load to the maximum load. Chiang and Jean-Jumeau [4] developed a performance index which gives the relationship between the value of index and the maximum load demand that the power system can withstand without collapse. Voltage stability factor (L_{mn}) has been proposed by Moghavvemi and Omar [5] to investigate the stability of each line of the system by calculating line stability indices. Musirin and Rahman [6] detected Fast Voltage Stability Index (FVSI) to detect the instability of the system for line outage. Gu and Wan [7] developed an index to use the measurements of the PMU for the calculation of an index to assess large system voltage stability. Devaraj and Roselyn [8] proposed ANN based on-line voltage stability assessment. Rajive and Tiwari [9] developed a new index LCPI based upon the entire model and ABCD parameters of the transmission network.

Kanimozhi and Selvi [10] developed a New Voltage Stability Index (NVSI) to find voltage instability of the system and it considers the effect of real power and reactive power loading of the system. Perez-londono et al. [11] proposed a new measurement based simplified voltage stability index (SVSI) that can enhance the online voltage monitoring by reducing the computational times. Mishra et al. [12] discussed the composite severity index (CSI), that combines the active power performance index and line stability index. Danish [13] proposed a new voltage stability margin (VSM) index using synchro phasor data. Chen and Jiang [14] proposed a wide area measurement-based voltage stability index which depends on the sensitivity of the L- index. Saurabh et al. [15] proposed line voltage stability index (LVSI) to determine the stability of lines by considering the ABCD parameters, resistance and line charging capacitance of the line. Li et al. [16] developed a VVC strategy by considering voltage deviation and SVS steady state voltage stability to coordinate VVC devices in PV clusters. Danish et al. [17] addressed review of voltage indices and their comparative analysis. Nian Zhang [18] discussed different types of Gaussian Process Regression methods. Wehenkel [19-20] proposed various machine learning methods for security assessment.

Chiang H.D et al. [21] presented a tool called CPFLOW, to trace the power system steady state behaviour with variations of parameters. Katsanevakis et al. [22] developed a voltage stability quality index (VSQI) to express as to how the operations of energy storage systems affect the voltage stability and quality of the local power system. Villa-Acevedo et al. [23] proposed a novel approach, kernel extreme learning machine and a mean variance mapping optimization (KELM-MVMO) algorithm to estimate the voltage stability margin. Many voltage stability indices are developed [24-30] for the assessment of stability state of the system and to identify the critical line of the transmission lines, maximum loading at bus in the system and then the weakest bus. Zhao et al. [31] developed a "Learning-to-Infer" algorithm to predict the status of every line in the network. Farsi et al. [32] discussed various algorithms to analyze their performance for short-term load forecasting. Dhibi et al. [33] proposed two enhanced RF classifiers and K-means clustering based reduced kernel RF (RK-RF K-means), for fault detection and diagnosis. (FDD) [34-36] proposed various machine learning algorithms for interval prediction of onboard solar power, for optimal power flow and for unbalanced distribution systems. Li et al. [37] developed a TSA methodology based on extreme gradient boosting (XGBoost) and factorization machine (FM). Hong et al. [38] presented a new approach to state estimation (SE) of distribution networks.

A thorough review of the available literature motivated the author to develop an alternative index, Modified Voltage Stability Index (MVSI) to assess voltage stability. This index considered the active and reactive power of the transmission line, sending end voltage (P_j, Q_j and V_i) and transmission line ABCD parameters to assess the stability of voltage. The relative directions of real and reactive power flows are also considered for stability analysis. Using NR power flow solution, all the required parameters are obtained for the assessment of voltage stability. Therefore, the proposed index provides an alternate method to give an accurate and faster assessment for voltage stability.

In this work, an applied soft computing technique is also incorporated for online assessment of voltage stability. This method enhances the assessment of voltage stability.

The Existing Indices for Assessment of Voltage Stability

Existing voltage indices L_{mn}, FVSI, NVSI and proposed index MVSI are used to assess the stability state of the power system shown in Figure 1. The proposed index MVSI results are compared with the results of the existing indices L_{mn}, FVSI and NVSI. These L_{mn}, FVSI and NVSI indices are discussed in the subsequent sections.

Line Stability Index (L_{mn})

L_{mn} [5] is calculated for each transmission line of a power system which is connected between any two buses. If the value of L_{mn} of all the transmission lines is less than one, the system is under stable condition. When the index of any transmission line is nearer to one, the system is close to getting unstable.

$$L_{mn} = \frac{4Q_j X}{[V_i \sin(\theta-\delta)]^2} \tag{1}$$

In Figure 1, V_i is voltage of bus i, X is the reactance of line, Q_j is the reactive power at bus j, $\delta = \delta_i - \delta_j$ is the phase difference between the phase angles of bus i and bus j, θ is the impedance angle of the line.

Fast Voltage Stability Index (FVSI)

The Fast Voltage Stability Index, FVSI [6], is defined as:

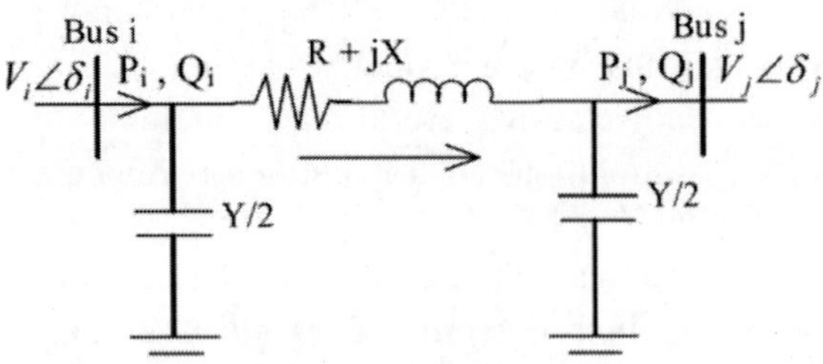

Figure 1. Typical two bus transmission line diagram.

$$FVSI = \frac{4Z^2 Q_j}{V_i^2 X} \tag{2}$$

where Z is the impedance of the line.

If the value of FVSI is close to unity, it indicates that the particular transmission line is close to instability. To maintain the stable condition of the system, the value of FVSI should be less than one.

New Voltage Stability Index (NVSI)

The New voltage stability index has been developed by Kanimozhi [10]. In the calculation of this stability index, active and reactive powers of transmission lines are considered and the resistance, shunt admittance of transmission lines are neglected.

$$NVSI = \frac{2X\sqrt{P_j{}^2+Q_j{}^2}}{2Q_jX-V_i{}^2} \tag{3}$$

The value of NVSI of all transmission lines must be less than 1 to maintain the system in stable condition.

Proposed Modified Voltage Stability Index (MVSI)

The Modified voltage stability index (MVSI) method developed by considering the entire model of transmission line and the real and reactive power flows in the line. The pie model of a transmission line system with two buses is shown in Figure 1. TheABCD line parameters and the sending end and the receiving end voltages can be expressed as

$$\begin{bmatrix} V_i \\ I_i \end{bmatrix} = \begin{bmatrix} A & B \\ C & D \end{bmatrix} \begin{bmatrix} V_j \\ I_j \end{bmatrix} \tag{4}$$

$$A = \left[1 + \frac{YZ}{2}\right] = \mid A \mid \angle\alpha \tag{5}$$

$$B = Z = \mid B \mid \angle\beta \tag{6}$$

$$C = Y\left[1 + \frac{YZ}{4}\right] \tag{7}$$

$$D = \left[1 + \frac{YZ}{2}\right] \tag{8}$$

where Z is the impedance and Y is the line charging admittance of the transmission line, respectively.

The complex power flow equation at receiving end is given by

$$P_j + jQ_j = \frac{|V_i||V_j|}{|B|}\angle\beta - \delta - \frac{|A||V_j{}^2|}{|B|}\angle\beta - \alpha \tag{9}$$

The real part of the above equation,

$$P_j = \frac{|V_i||V_j|}{|B|}cos\,(\beta - \delta) - \frac{|A||V_j{}^2|}{|B|}cos\,(\beta - \alpha) \tag{10}$$

And imaginary part,

$$Q_j = \frac{|V_i||V_j|}{|B|}sin\,(\beta - \delta) - \frac{|A||V_j{}^2|}{|B|}sin\,(\beta - \alpha) \tag{11}$$

From Eqs. (10) and (11)

$$cos\,(\beta - \delta) = [P_j + \frac{|A|}{|B|} \mid V_j^2 \mid cos\,(\beta - \alpha)]\frac{|B|}{|V_i||V_j|} \tag{12}$$

$$sin\,(\beta - \delta) = [Q_j + \frac{|A|}{|B|} \mid V_j^2 \mid sin\,(\beta - \alpha)]\frac{|B|}{|V_i||V_j|} \tag{13}$$

Since $cos^2\,\theta + sin^2\,\theta = 1$

Then employing the trigonometry identity given by

$$cos^2\,(\beta - \delta) + sin^2\,(\beta - \delta) = 1$$

$$\{P_j + \frac{|A|}{|B|} \mid V_j^2 |cos\,(\beta - \alpha)\ \frac{|B|}{|V_i||V_j|}\}^2 + \{Q_j + \frac{|A|}{|B|} \mid V_j^2 \mid sin(\beta - \alpha)\frac{|B|}{|V_i||V_j|}\}^2 = 1 \tag{14}$$

After simplification

$$\mid V_j^4 \mid + \frac{2B}{A}\left[P_j{}^2 cos\,(\beta - \alpha) + Q_j{}^2 sin\,(\beta - \alpha) - \frac{V_i{}^2}{A^2}\right] + \frac{[P_j{}^2 + Q_j{}^2]B^2}{A^2} = 0 \tag{15}$$

For real solution of Eqs. (15), the discriminate must be greater than or equal to zero. i.e., $(b^2\text{-}4ac) \geq 0$

$$\frac{\sqrt{[P_j^{\,2}+Q_j^{\,2}]}}{\left[P_j cos\,(\beta-\alpha)+Q_j sin\,(\beta-\alpha)-\frac{V_i^{\,2}}{2\,AB}\right]} \leq 1 \qquad (16)$$

The proposed index value should be less than one in order to make sure that the system is stable and secure.

$$MVSI = \frac{\sqrt{[P_j^{\,2}+Q_j^{\,2}]}}{\left[P_j cos\,(\beta-\alpha)+Q_j sin\,(\beta-\alpha)-\frac{V_i^{\,2}}{2\,AB}\right]} \qquad (17)$$

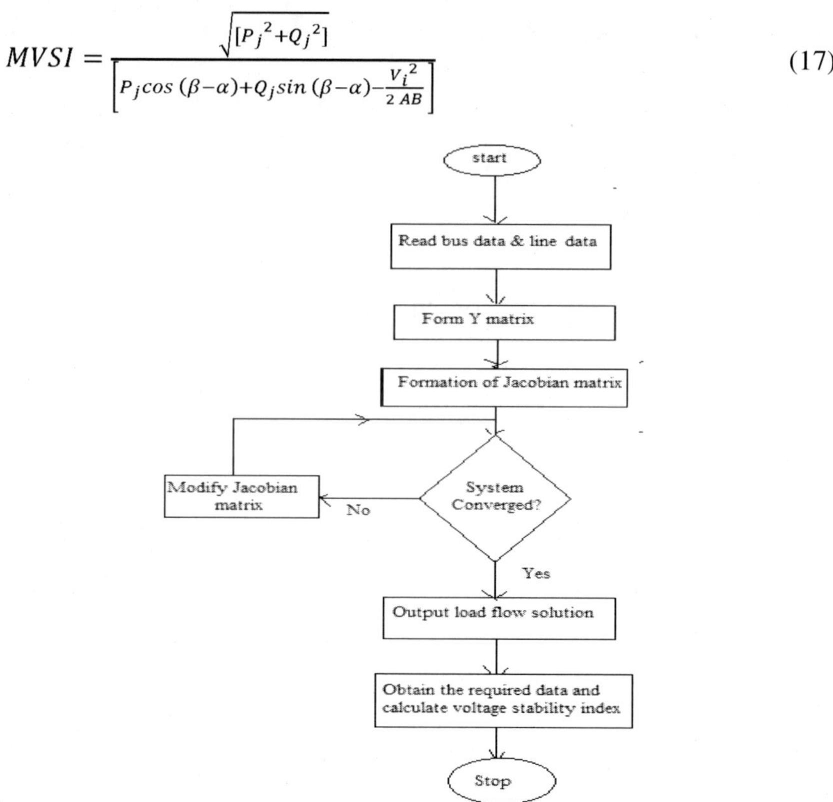

Figure 2. The flow chart of the N R Iterative Method for the voltage stability index.

From Eq. (17), it is identified that the system is unstable when the MVSI is closer to one. To maintain the secure condition of the power system, the value of MVSI should be less than one. The NR power flow method is used for obtaining the values of real and reactive power flows. The Flow chart for the Newton-Raphson iterative method and the VSI calculation is as shown in Figure 2.

Results and Comparative Analysis of MVSI vs Other Indices

The proposed MVSI tested on IEEE 30 bus system and the comparison of the proposed MVSI with the other voltage indices L_{mn}, FVSI and NVSI is given in Table 1-4.

IEEE 30 Bus System Results

Base Load Operating Condition

The Proposed MVSI is compared with other indices L_{mn}, FVSI and NVSI at base loading condition. Table 1 shows the calculated values of different voltage indices at base load. From Table1, it is identified that the L_{mn} and FVSI indices are lower than the MVSI for the lines 5-7, 8 -28, 10-21, 10-22, 25-26 and for the lines 4-6, 12-14, 15-18, 27-29, 27-30 and 29-30, the other indices L_{mn}, FVSI are higher than the proposed index. This may be due to the directions and magnitudes of real and reactive power flows. If the direction of real and reactive powers is the same, the value of the proposed index is higher than the other indices. If the direction of real and reactive powers is opposite, the value of the proposed index is lower than the other indices. But the index value of NVSI [10] is similar to the proposed index MVSI for all the lines.

Table 1. Indices of the IEEE 30 bus system at base loading

Line from - to	Proposed index (MVSI)	NVSI	L_{mn}	FVSI
4 to 6	0.0608	0.0576	0.0312	0.0306
5 to 7	0.0436	0.041	0.0544	0.0553
8 to 28	0.0101	0.0095	0.0204	0.0205
10 to 21	0.0287	0.0258	0.0331	0.0329
10 to 22	0.0274	0.0245	0.0307	0.0305
12 to 14	0.0428	0.038	0.0279	0.0275
15 to 18	0.0288	0.0256	0.0178	0.0176
24 to 25	0.0152	0.0132	0.0266	0.0268
25 to 26	0.0387	0.0317	0.0507	0.0502
27 to 29	0.0592	0.0512	0.0345	0.0337
27 to 30	0.0997	0.0848	0.0508	0.0488
29 to 30	0.0388	0.0338	0.0141	0.0139

This is due to the consideration of line parameters and the real and reactive power flows in the NVSI index like in MVSI.

Heavy Active Loading Condition

For an IEEE 30 bus system, at each bus, active power loading is increased continuously till the voltage collapse occurs in the system. The effect of active power loading of the proposed index and other indices is shown in Table 2. And from Table 2, it is identified that the proposed index gives the condition of voltage collapse at an earlier stage, where the other indices are inadequate to identify the critical point. For the active power loading factor 2.11, the proposed index value is 1.0242, whereas other indices values are 0.8084, 0.4763 and 0.3476, which are extremely far from identifying the critical point of the power system. From table 2, it is also known as to which line and which bus are critical under increment of active power loading at each bus for the proposed index.

Table 2. Active power loading at a single bus of IEEE 30-bus system

Load bus	Critical line	Loading factor	Proposed index (MVSI)	NVSI	L_{mn}	FVSI
3	1 to 3	2.11	1.0242	0.8084	0.4763	0.3476
4	1 to 3	2.54	0.9956	0.7899	0.4577	0.3369
10	6 to 10	2.10	0.9984	0.9985	0.6592	0.5557
12	4 to 12	1.568	0.9916	0.9917	0.5439	0.4769
14	4 to 12	1.25	0.9986	0.9986	0.6380	0.5641
16	4 to 12	1.592	0.9902	0.9902	0.5859	0.5142
23	4 to 12	1.31	1.0006	1.0006	0.6624	0.5842
26	25 to 26	0.322	0.9823	0.9876	1.2924	1.0186
30	27 to 30	0.507	0.9641	0.9544	1.4943	0.7970

Heavy Reactive Loading Condition

Table 3 gives the results of indices for incrementing reactive load continuously at each bus of IEEE 30 bus system till voltage collapse occurs in the system. From the results of reactive power increment at each bus, it is identified that the proposed index is also able to identify the instability of the system as well as other indices NVSI, L_{mn}, FVSI. But the voltage indices L_{mn} and FVSI are giving the condition of voltage collapse at an earlier stage as these are more dependent on reactive power loading. Thus, for all the reactive power loading factors, the values of L_{mn} and FVSI are higher than the proposed index (MVSI). The values of the proposed index and NVSI index are approximately the same because of the consideration of line parameters and real and reactive power flows.

Table 3. Reactive power loading at a single bus of IEEE 30-bus system

Load bus	Critical line	Loading factor	Proposed index (MVSI)	NVSI	L_{mn}	FVSI
3	1 to 3	2.62	0.9627	0.9628	0.9738	0.9650
4	2 to 4	4.768	0.9905	0.9899	1.0073	0.9800
10	6 to 10	1.9	1.0067	1.0068	1.0765	1.0687
12	12 to 13	2.325	1.0221	1.0206	1.1768	1.1768
14	14 to 15	1.04	1.0084	0.9986	1.1379	1.2651
16	16 to 17	1.33	0.9909	0.9952	1.1721	1.2469
23	23 to 24	0.96	1.0351	1.0006	1.2465	1.2939
26	25 to 26	0.422	1.0235	0.9953	1.2161	1.2319
30	27 to 30	0.34	1.0650	0.9943	1.2665	1.3330

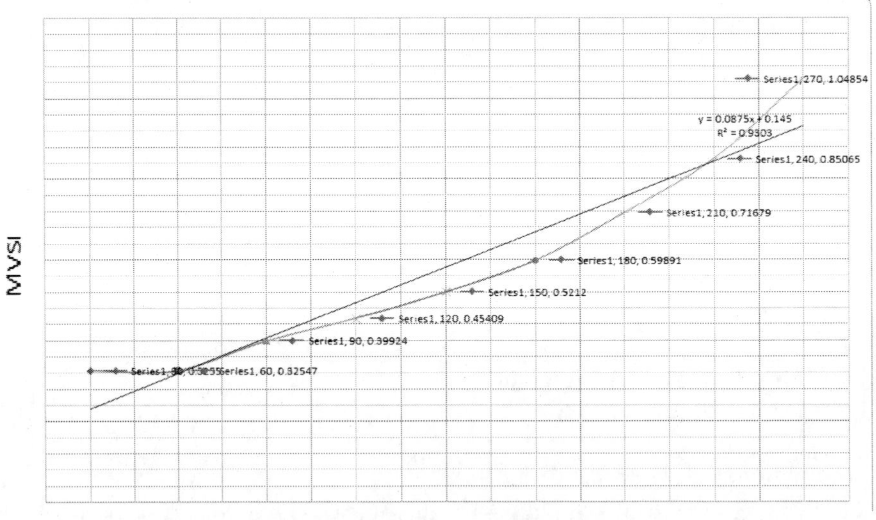

Figure 3. Reactive loading at bus 3 of line 1-3 with MVSI.

From Table 3, it is evident that by increasing reactive power at bus 3 up to 2.62 p.u or 262MW in the line 1 to 3 proposed index is 0.99276, which is more nearer to one as compared with the other bus at this condition of loading. Thus, it can be treated as a more critically stressed line (i.e., beyond the increase in load, it will become unstable). Figure 3 shows MVSI variation with respect to a gradual increase in reactive power at bus 3. The proposed index can be able to identify the more stressed lines, during heavy active and reactive loading conditions. It shows that MVSI is more suitable for digital simulation procedures to predict voltage stability using soft computing techniques.

Heavy MVA Loading Condition

Table 4 shows the results of values of indices for increasing both active power and reactive power simultaneously at bus 3. At the loading factor 2.07, the proposed index MVSI is 1.00234, which shows the unstable condition of the system. It also shows that the values of other indices are far from the critical state values.

By applying various loading conditions to IEEE 30 bus system, the proposed index is calculated and compared with different valid stability indices. From the results, it is concluded that the proposed method is more effective and accurate.

Table 4. Real and reactive power loading at a single bus of IEEE 30-bus system. (MVA loading)

Loading factor	Critical line	Proposed index (MVSI)	NVSI	L_{mn}	FVSI
2.07	1 to 3	1.0023	0.7947	0.4788	0.3533

Table 5. Indices of the IEEE 57 bus system at base loading

Line from – to	Proposed index (MVSI)	NVSI	L_{mn}	FVSI
3 to 4	0.0902	0.0980	0.0196	0.0193
6 to 8	0.2001	0.2017	0.0177	0.0181
9 to 11	0.0659	0.0642	0.0083	0.0082
12 to 13	0.0576	0.5561	0.1114	0.1120
21 to 20	0.010	0.0100	0.0100	0.0100
24 to 26	0.0017	0.0016	0.0017	0.0017
7to 29	0.0308	0.0308	0.0308	0.0308
30 to 31	0.0611	0.0581	0.0807	0.0795
35 to 36	0.0210	0.0212	0.0231	0.0233
37 to 38	0.0576	0.0554	0.0764	0.0776
41 to 42	0.0801	0.0761	0.0617	0.0599
38 to 44	0.0309	0.2905	0.0149	0.0151
13 to 49	0.2080	0.2062	0.2622	0.2614
53 to 54	0.0538	0.0513	0.0750	0.07611
56 to 41	0.0848	0.0864	0.0312	0.0334
9 to 55	0.0762	0.0756	0.0526	0.0525
39 to 57	0.1034	0.1010	0.1622	0.1617

IEEE 57 Bus System Results

Base Load Operating Condition

The Proposed MVSI is compared with other indices L_{mn}, FVSI and NVSI at base loading condition. Table 5 shows the calculated values of different voltage indices at base load. From the table 5 it is identified that the L_{mn} and FVSI indices are lower than the MVSI for the lines 3-4, 6-8, 9 -11, 41-42, 38-44, 56-41 and for the lines 12-13, 30-31, 35-36, 37-38, 13-49, 53-54, and 39-57, the other indices L_{mn}, FVSI are higher than the proposed index.

Active Power Loading Condition

For an IEEE 57 bus system active power loading is increased continuously at a few selected buses till the voltage collapse occurs in the system. The effect of active power loading of the proposed index and other indices are shown in Table 6. From Table 6, it is identified that the proposed index gives the condition of voltage collapse at the earlier stage, where the other indices are inadequate to identify the critical point. For the active power loading factor 4.68 at load bus 5, the proposed index value is 0.9768, whereas other indices values are 0.9512, 0.6607 and 0.6565, which are very far from identifying the critical point of the power system. Further, from Table 6, it is also known that which line and which bus is critical under increment of active power loading at each bus for the proposed index.

Table 6. Active power loading at a single bus of IEEE 57-bus system

Load bus	Critical line	Loading factor	Proposed index (MVSI)	NVSI	L_{mn}	FVSI
5	4 to 5	4.68	0.9768	0.9512	0.6607	0.6565
15	1 to 15	4.576	0.9909	0.9499	0.6552	0.6361
29	7 to 29	2.478	0.9677	0.9526	0.7623	0.7431
33	34 to 32	0.297	0.9957	0.9752	0.8387	0.7414
38	13 to 49	3.26	0.9979	0.9723	0.6144	0.4721
41	11 to 41	0.976	0.9903	0.9827	0.7126	0.6328
49	13 to 49	2.628	0.9921	0.9782	0.8468	0.8400
52	29 to 52	1.227	0.9990	0.9895	0.6847	0.6408
55	9 to 55	2.828	0.9924	0.9986	0.5638	0.5521

Reactive Power Loading Condition

Table 7 gives the results of indices for incrementing reactive load continuously at each bus of IEEE 57 bus system till voltage collapse occurs in the system.

From the results of reactive power increment at each bus, it is identified that the proposed index is also able to identify the instability of the system as well as other indices NVSI, L_{mn}, FVSI. But the voltage indices L_{mn} and FVSI are giving the condition of voltage collapse at the earlier stage as these are more dependent on reactive power loading. So, for all the reactive power loading factors the values of L_{mn}, FVSI are higher than the proposed index (MVSI). Also, the values of the proposed index and NVSI index are approximately the same because of the consideration of line parameters and real and reactive power flows.

Table 7. Reactive power loading at a single bus of IEEE 57-bus system

Load bus	Critical line	Loading factor	Proposed index (MVSI)	NVSI	L_{mn}	FVSI
5	4 to 5	3.6	0.9600	0.9545	1.0428	0.9656
15	1 to 15	5.775	0.9509	0.9489	0.9949	0.9973
29	7 to 29	2.25	0.9785	0.9627	0.9999	0.9998
33	34 to 32	0.20	0.8951	0.8875	0.9999	0.9997
38	13 to 49	2.31	0.9832	0.9781	1.0215	1.0154
41	11 to 41	0.618	0.9897	0.9672	1.0422	1.0366
49	13 to 49	1.785	0.9825	0.9743	1.0456	1.0449
52	29 to 52	0.948	0.9777	0.9525	1.1099	1.0014
55	9 to 55	1.564	0.9968	0.9873	1.0486	1.0488

IEEE 118 Bus System Results

Base Load Operating Condition

The Proposed MVSI is compared with other indices L_{mn}, FVSI and NVSI at base loading condition. Table 8 shows the calculated values of different voltage indices at base load. From the table 8 it is identified that the L_{mn} and FVSI indices are higher than the MVSI for the lines 77-69, 96-80, 97-80, 98-80, 99-80 and for the lines 3-12, 13-11, 62-66 the other indices L_{mn}, FVSI are lower than the proposed index. For the lines 25-26, 37-38, 66-65, 69-68 and 80-81 all indices are the same.

Active Power Loading Condition

For an IEEE 118 bus system active power loading is increased continuously at a few selected buses till the voltage collapse occurs in the system. The effect of active power loading of the proposed index and other indices are shown in Table 9. From Table 9, it is identified that the proposed index gives the condition of voltage collapse at the earlier stage, where the other indices are inadequate to identify the critical point. For the active power loading factor 8.792 at bus 11, the proposed index value is 0.9988, whereas other indices values are 0.9752, 0.6077 and 0.5605, which are very far from the identifying critical point of the power system. Further, from table 9, it is also known that which line and which bus is critical under increment of active power loading at each bus for the proposed index.

Table 8. Indices of the IEEE 118 bus system at base loading

Line from – to	Proposed index (MVSI)	NVSI	L_{mn}	FVSI
3 to 12	0.0902	0.0880	0.0832	0.0837
5 to 8	0.2001	0.174	0.1315	0.1304
13 to 11	0.0659	0.0626	0.0598	0.0589
25 to 26	0.0320	0.0327	0.0321	0.0320
30 to 8	0.1310	0.1108	0.1920	0.1907
37 to 38	0.1250	0.1255	0.1173	0.1163
38 to 65	0.3742	0.3527	0.1093	0.1092
51 to 49	0.2243	0.2129	0.1611	0.1511
54 to 49	0.2715	0.2503	0.1939	0.1808
55 to 59	0.1607	0.1515	0.0555	0.0570
62 to 66	0.1645	0.1625	0.1284	0.1319
66 to 65	0.0624	0.0606	0.0609	0.0609
67 to 66	0.1149	0.1092	0.0882	0.0863
69 to 68	0.1461	0.1433	0.1457	0.1454
75 to 69	0.3076	0.2767	0.1608	0.1456
77 to 69	0.0316	0.0304	0.0523	0.0503
80 to 81	0.0764	0.0752	0.0772	0.0772
86 to 87	0.0608	0.0515	0.1239	0.1240
96 to 80	0.1300	0.1287	0.2053	0.2033
99 to 80	0.078	0.0772	0.0897	0.0882
94 to 100	0.0593	0.0522	0.1302	0.1293
97 to 80	0.0795	0.0752	0.1220	0.1211
98 to 80	0.0534	0.0527	0.0676	0.0668
106 to 105	0.0124	0.0118	0.0133	0.0133
114 to 32	0.0136	0.0103	0.0118	0.0118

Table 9. Active power loading at a single bus of IEEE118-bus system

Load bus	Critical line	Loading factor	Proposed index (MVSI)	NVSI	L_{mn}	FVSI
11	3 to 12	8.792	0.9988	0.9752	0.6077	0.5605
16	12 to 16	4.6	0.9954	0.9899	0.4577	0.4369
45	45 to 49	3.397	0.9987	0.9982	0.4623	0.4431
52	52 to 53	3.259	0.9401	0.927	0.4711	0.4877
101	101 to 102	4.622	0.9997	0.9986	0.5638	0.5521
108	103 to 110	4.327	0.9854	0.972	0.6599	0.5142
117	12 to 117	1.79	0.9984	0.9906	0.6144	0.4721

Reactive Power Loading Condition

Table 10 shows the results of indices for incrementing reactive load continuously at each bus of IEEE 118 bus system till voltage collapse occurs in the system. From the results of reactive power increment at each bus, it is identified that the proposed index is also able to identify the instability of the system as well as other indices NVSI, L_{mn}, FVSI. But the voltage indices L_{mn} and FVSI are giving the condition of voltage collapse at the earlier stage as these are more dependent on reactive power loading. So for all the reactive power loading factors the values of L_{mn}, FVSI are higher than the proposed index (MVSI). Also, the values of the proposed index and NVSI index are approximately the same because of the consideration of line parameters and real and reactive power flows.

Table 10. Reactive power loading at a single bus of IEEE118-bus system

Load bus	Critical line	Loading factor	Proposed index (MVSI)	NVSI	L_{mn}	FVSI
11	4 to 11	13.11	0.9999	0.9989	1.00410	1.01274
16	12 to 16	2.91	0.9946	0.9923	1.00701	1.02851
45	45 to 46	2.97	0.9947	0.9976	1.0052	1.0035
52	49 to 51	2.19	0.9975	0.9962	0.99224	0.95790
101	100 to 101	2.415	0.9995	0.9988	1.13118	1.16384
108	105 to 108	3.64	0.9989	0.9978	1.01684	1.05528
117	12 to 117	1.185	0.9956	0.9859	1.00283	1.01009

The Machine Learning Approach for Voltage Stability Assessment

Machine learning algorithms like ANN, decision tree, etc. [19-20] can be able to provide required security information of power systems. This work presents a novel machine learning approach for online voltage stability assessment of the power system. In this paper, the Gaussian process regression [GPR] machine learning algorithm [18] is used for online voltage stability index assessment. The GPR algorithm is non parametric supervised machine learning and is applicable for small data set. It is very useful for predicting uncertainty measurements. The Exponential GPR model is giving better results for the given data set when it is compared with other methods in terms of prediction accuracy, training time, prediction speed, Root mean square error [RMSE]. The proposed exponential GPR model is trained offline. The data set required to train the GPR model is obtained from the MATLAB simulation results using the NR load flow method.

The Exponential GPR Machine Learning Algorithm

The Exponential Gaussian Process Regression deals smooth functions with minor errors, but it does not handle discontinuous functions.

The Gaussian process regression algorithm:

Inputs:

1. The input data to the model for training is in the form of:

$$\{(x_{p,}y_{p}) ; p = 1,2, \ldots \ldots n\} \tag{18}$$

$where\ x_p \epsilon R^d and\ y_p \epsilon R$,

The GPR model predicts the new response value of y for the given new input x.

2. The representation of a linear regression model is:

$$y = X^T \beta + \varepsilon \tag{19}$$

Procedure:

1. Let the n point input data set be in the form of:

$$\{(x_{p,}y_{p}) ; p = 1,2, \ldots \ldots n\}$$

$where\ x_p \epsilon R^d and\ y_p \epsilon R$

2. The representation of a linear regression model is:

$$Y = X^T \beta + \varepsilon$$

3. The model of linear regression, where K(X, X) is parametrized and looks as:

$$K(X,X) = (K_1(x_1,x_1) \cdots K_1(x_1,x_n) \vdots \ddots \vdots K_n(x_n,x_1) \cdots K_n(x_n,x_n)) \quad (20)$$

4. The Exponential Gaussian Process Regression represented as:

$$K(x_i, x_j \mid \theta) = \sigma_f{}^2 \, exp \, exp \left[- \frac{r}{\sigma_l}\right] \quad (21)$$

$$where; \; r = \sqrt{(x_i - x_j)^T (x_i - x_j)}$$

Methodology

For training the Exponential GPR model, there are 4264 observations with two input variables and one output variable IEEE 30 bus system, 6720 observations for IEEE 57 bus and 6510 observations for IEEE 118 bus system. Active and reactive powers of lines at various loading conditions are used as input variables where MVSI is an output variable for the proposed GPR model. The GPR model is trained offline. After training, the trained model can be able to predict the value of MVSI for any given unknown loading condition. To test the performance of the model, the input variables are given to the model at base load, 1.5 load factor and 2.5 loading factor for testing IEEE 30 bus system and input variables are given at base load, 0.5 load factor and 1.5 load factor for IEEE 57 bus and 0.85 load factor and 1.15 loading factor for IEEE 118 bus system. The modified voltage stability index (MVSI) is evaluated at different loading conditions. The comparison of the results of the trained model and MATLAB simulations using the NR method is shown in Table 11 for IEEE 30 bus system Table 12 for IEEE 57 bus system and Table 13 for IEEE 118 bus system.

Results and Comparative Analysis of Exponential GPR vs NR Method MVSI Indices

Comparative Analysis

IEEE 30 Bus System

Using the trained exponential GPR model, modified voltage stability index (MVSI) values are predicted for the given input variables at base loads and at

MVA loading factor 1.5 and 2.5 of IEEE 30 bus data. The predicted index values by this machine learning algorithm are compared with the true index values of the MATLAB simulation. The comparison is shown in Table 11, The Exponential GPR algorithm giving the same results at the base load and also at 1.5 times the base load and 2.5 times the base load (i.e., P&Q loading).

The maximum error between the predicted values and the true values is approximately 0.01. Results in Table 4 shows that, the line 1-3 is critically stressed at 2.07 times MVA loading where the MVSI is 1.00234; but, with a further increase in the load, the line will become unstable.

When the proposed GPR algorithm is tested at 1.5-times MVA loading, the line 1-3 is more stressed (i.e., index value is 0.5774) but not unstable. But at 2.50 times MVA loading the line 1-3 is critically stressed (i.e., index value is 1.5793) and becomes unstable. The GPR algorithm is tested at the base, under and overloading for IEEE 30 bus test system.

IEEE 57 Bus System

Using the trained exponential GPR model, modified voltage stability index (MVSI) values are predicted for the given input variables at base loads and at MVA loading factor 0.5 and 1.5 of IEEE 57 bus data. The predicted index values by this machine learning algorithm are compared with the true index values of the MATLAB simulation. The comparison is shown in Table 12, The Exponential GPR algorithm gives the results with maximum error 0.0245 at the base load and at 0.5 times the base load and 1.5 times the base load (i.e., P&Q loading).

The maximum error between the predicted values and the true values is approximately 0.0245. Results in Table 11 show that the lines 1-15 and 1-16 are critically stressed at 1.5 times MVA loading where the MVSI is 1.0952 and 1.0113, but, with a further increase in the load, the line will become unstable.

When the proposed GPR algorithm is tested at 1.5-times MVA loading, the line 1-15 and 1-16 are more stressed (i.e., index values are 1.0967 and 1.028) and become unstable. The GPR algorithm is tested at the base, under and over loading for IEEE 57 bus test system.

IEEE 118 Bus System

Using the trained exponential GPR model, modified voltage stability index (MVSI) values are predicted for the given input variables at base loads and at MVA loading factor 0.85 and 1.15 of IEEE 118 bus data. The predicted index values by this machine learning algorithm are compared with the true index values of the MATLAB simulation. The comparison is shown in Table 13, The Exponential GPR algorithm gives the results with maximum error 0.025 at the base load and at 0.85 times the base load and 1.15 times the base load (i.e., P&Q loading).

Table 13 shows the results of the true index values of the MATLAB simulation and the proposed GPR algorithm tested at 15% below and 15% above the base load condition for IEEE 118 bus system. The IEEE 118 system is overloading for 1.16 times MVA loading, the line 1-2 is more stressed (i.e., index value is 0.9955) near to unstable. So that at 1.17 times MVA loading the line 1-2 is critically stressed (i.e., index value is 1.1716) and becomes unstable. The GPR algorithm is tested at the base, under and overloading for IEEE 118 test system. The Gaussian process regression [GPR] machine learning algorithm results are consistent with MATLAB results. This evolution shows that Machine learning algorithms can be used for online assessment of the voltage stability of the power system.

Conclusion

This work presents an alternative new index MVSI for voltage stability assessment of a power system. An IEEE 30 bus system, IEEE 57 bus system and IEEE 118 bus systems are used for testing the proposed MVSI under various possible active power loading, reactive power loading and MVA loading conditions. From the results it is observed that for active power loading and MVA loading the proposed index gives the condition of voltage collapse at an earlier stage, where the other indices are inadequate to identify the critical point. For reactive power loading it is also able to identify the instability of the system like other indices NVSI, L_{mn}, FVSI which are more dependent on reactive power loading. And results are satisfactory when compared with the other existing indices. The present work also demonstrated the machine learning approach for assessment of voltage stability. The Gaussian process regression machine learning algorithm is used for online voltage stability index assessment. It shows that an exponential GPR model is giving better results compared to the other machine learning algorithms available. The predicted results of a trained exponential GPR model are compared with the results of MATLAB simulated NR method. The results are consistent and converged. It is evident that Machine learning algorithms can be used for the online assessment of voltage stability in the power system network.

Table 11. Comparison of MVSI of GPR and NR methods at various loading conditions of IEEE 30 bus system

S. No	Line From - To	Base load		At MVA loading 1.5 times of the base load		At MVA loading 2.5 times of the base load	
		Index by NR method [MVSI]	Index by Machine learning [Exponential GPR method]	Index by NR method [MVSI]	Index by Machine learning [Exponential GPR method]	Index by NR method [MVSI]	Index by Machine learning [Exponential GPR method]
1.	1-2	0.2007	0.2005	0.3702	0.3714	0.835	0.845
2.	1-3	0.3076	0.3076	**0.5774**	**0.5775**	**1.5782**	**1.5793**
3.	2-4	0.1624	0.1617	0.2743	0.2739	0.5631	0.5623
4.	3-4	0.0611	0.0625	0.1066	0.1086	0.2143	0.2165
5.	2-5	0.3355	0.335	0.5969	0.5967	1.2128	1.2087
6.	2-6	0.2245	0.2248	0.3884	0.3863	0.7921	0.7948
7.	4-6	0.0608	0.0611	0.104	0.1043	0.2209	0.2218
8.	5-7	0.0436	0.0437	0.0818	0.0818	0.1737	0.1698
9.	6-7	0.0642	0.0644	0.1094	0.1103	0.1953	0.1923
10.	6-8	0.0255	0.0257	0.047	0.0471	0.1639	0.1623
11.	6-9	0.1196	0.1187	0.2056	0.2053	0.3697	0.3695
12.	6-10	0.1731	0.1726	0.2977	0.2965	0.5981	0.5911
13.	9-11	0.0621	0.0616	0.1701	0.1687	0.2883	0.2796
14.	9-10	0.0574	0.0625	0.1078	0.1121	0.2251	0.2283
15.	4-12	0.2491	0.2505	0.4145	0.4145	0.8289	0.8289
16.	12-13	0.0264	0.0302	0.0802	0.0805	0.2062	0.2063
17.	12-14	0.0428	0.0441	0.0735	0.0808	0.1443	0.1536
18.	12-15	0.0518	0.0534	0.0896	0.0917	0.1739	0.1839
19.	12-16	0.032	0.0322	0.0547	0.0547	0.1072	0.1072
20.	14-15	0.0095	0.0088	0.0173	0.0174	0.039	0.039

S. No	Line From - To	Base load		At MVA loading 1.5 times of the base load		At MVA loading 2.5 times of the base load	
		Index by NR method [MVSI]	Index by Machine learning [Exponential GPR method]	Index by NR method [MVSI]	Index by Machine learning [Exponential GPR method]	Index by NR method [MVSI]	Index by Machine learning [Exponential GPR method]
21.	16-17	0.0152	0.0131	0.0264	0.0262	0.055	0.0567
22.	15-18	0.0288	0.0306	0.0499	0.074	0.1042	0.1143
23.	18-19	0.0079	0.0106	0.0138	0.0159	0.0302	0.0312
24.	19-20	0.0105	0.0105	0.0184	0.0182	0.0411	0.0471
25.	10-20	0.0418	0.042	0.0723	0.072	0.1474	0.1465
26.	10-17	0.0116	0.0117	0.0196	0.0217	0.04	0.04
27.	10-21	0.0287	0.0273	0.0506	0.0519	0.1054	0.1104
28.	10-22	0.0274	0.029	0.0489	0.0495	0.1028	0.1082
29.	21-22	0.0012	0.0013	0.0017	0.0017	0.0029	0.0027
30.	15-23	0.0247	0.028	0.0455	0.0453	0.0995	0.0984
31.	22-24	0.0256	0.0256	0.0497	0.0525	0.1135	0.1196
32.	23-24	0.0126	0.0138	0.0258	0.026	0.0633	0.0673
33.	24-25	0.0152	0.015	0.0233	0.0263	0.0327	0.0389
34.	25-26	0.0387	0.0374	0.0702	0.0699	0.1583	0.1482
35.	25-27	0.0227	0.0227	0.0385	0.0384	0.0855	0.0831
36.	28-27	0.1531	0.1556	0.2751	0.2736	0.57	0.5688
37.	27-29	0.0592	0.0507	0.1077	0.1070	0.2393	0.2289
38.	27-30	0.0997	0.1004	0.1826	0.1847	0.3761	0.3706
39.	29-30	0.0388	0.0397	0.0716	0.0688	0.1774	0.1678
40.	8-28	0.0101	0.0101	0.0098	0.0129	0.1234	0.1312
41.	6-28	0.0257	0.0276	0.0392	0.0393	0.0723	0.0745

Table 12. Comparison of MVSI of GPR and NR methods at various loading conditions of IEEE 57 bus system

S. No.	Line From – To	Base load		At MVA loading 0.5 times of the base load		At MVA loading 1.5 times of the base load	
		Index by NR method [MVSI]	Index by Machine learning [Exponential GPR method]	Index by NR method [MVSI]	Index by Machine learning [Exponential GPR method]	Index by NR method [MVSI]	Index by Machine learning [Exponential GPR method]
1.	3-4	0.0970	0.0998	0.0445	0.0453	0.2746	0.2767
2.	6-8	0.2001	0.2023	0.2384	0.2336	0.1115	0.1117
3.	9-11	0.0659	0.0554	0.0907	0.0874	0.0887	0.0753
4.	1-15	0.2806	0.2833	0.1276	0.1178	1.0952	1.0967
5.	1-16	0.3225	0.3192	0.1458	0.1461	1.0133	1.028
6.	12-13	0.0576	0.0577	0.0423	0.0463	0.1538	0.1538
7.	21-20	0.0105	0.0101	0.0168	0.0175	0.0155	0.0135
8.	24-26	0.0017	0.0051	0.015	0.016	0.0208	0.0399
9.	7-29	0.0308	0.0282	0.0542	0.0518	0.1345	0.144
10.	30-31	0.0611	0.0583	0.0327	0.024	0.1414	0.1264
11.	35-36	0.0211	0.0221	0.0078	0.0179	0.0501	0.0533
12.	37-38	0.0577	0.0577	0.0254	0.0366	0.1274	0.1126
13.	41-42	0.0802	0.0704	0.0435	0.0594	0.157	0.1371
14.	38-44	0.0309	0.0304	0.0076	0.0042	0.0882	0.106
15.	13-49	0.208	0.2068	0.1175	0.1083	0.3391	0.3275
16.	53-54	0.0538	0.0537	0.025	0.0204	0.0979	0.0911
17.	56-41	0.0848	0.0869	0.0622	0.0533	0.1792	0.1613
18.	9-55	0.0762	0.0742	0.0253	0.0452	0.0934	0.078
19.	39-57	0.1035	0.1011	0.0722	0.066	0.2844	0.2626

Table 13. Comparison of MVSI of GPR and NR methods at various loading conditions of IEEE 118 bus system

S. No.	Line From – To	Base load		At MVA loading 0.85 times of the base load		At MVA loading 1.15 times of the base load		At MVA loading 1.17 times of the base load	
		Index by NR method [MVSI]	Index by Machine learning [Exponential GPR method]	Index by NR method [MVSI]	Index by Machine learning [Exponential GPR method]	Index by NR method [MVSI]	Index by Machine learning [Exponential GPR method]	Index by NR method [MVSI]	Index by Machine learning [Exponential GPR method]
1.	1-2	0.0366	0.0384	0.5584	0.557	0.8533	0.845	1.1717	1.15048
2.	3-12	0.09027	0.0908	0.4732	0.4694	0.6579	0.6548	0.8312	0.8157
3.	8-5	0.2001	0.202	0.3468	0.3617	0.0681	0.0628	0.0688	0.0616
4.	11-13	0.0659	0.0754	0.0776	0.0754	0.2067	0.2066	0.2271	0.2537
5.	26-25	0.03206	0.0327	0.0295	0.0186	0.123	0.1507	0.1316	0.1256
6.	8-30	0.131	0.131	0.2602	0.2491	0.4879	0.4677	0.5265	0.5301
7.	38-37	0.125	0.1278	0.185	0.204	0.2512	0.2847	0.2565	0.238
8.	38-65	0.3742	0.3709	0.9299	0.9456	0.2493	0.2051	0.3605	0.3414
9.	49-51	0.2243	0.2228	0.1448	0.1369	0.215	0.2073	0.2142	0.1958
10.	49-54	0.2716	0.2536	0.1649	0.1501	0.261	0.2771	0.2603	0.1649
11.	55-59	0.1607	0.1598	0.1701	0.1694	0.2022	0.2127	0.1969	0.1823
12.	62-66	0.1645	0.1636	0.1663	0.1517	0.2236	0.2182	0.2349	0.2128
13.	65-66	0.0624	0.0684	0.0099	0.0199	0.0411	0.0455	0.0428	0.0542
14.	66-67	0.1149	0.1144	0.0869	0.0843	0.1224	0.1182	0.1222	0.111
15.	68-69	0.1461	0.1422	0.1133	0.1136	0.1069	0.1066	0.104	0.1023
16.	69-75	0.3077	0.3073	0.2317	0.2321	0.2662	0.2579	0.2551	0.2796
17.	69-77	0.0316	0.0453	0.0536	0.049	0.2188	0.1973	0.226	0.219
18.	80-81	0.0764	0.0763	0.1186	0.1234	0.065	0.0791	0.0745	0.0726

Table 13. (Continued)

S. No.	Line From – To	Base load		At MVA loading 0.85 times of the base load		At MVA loading 1.15 times of the base load		At MVA loading 1.17 times of the base load	
		Index by NR method [MVSI]	Index by Machine learning [Exponential GPR method]	Index by NR method [MVSI]	Index by Machine learning [Exponential GPR method]	Index by NR method [MVSI]	Index by Machine learning [Exponential GPR method]	Index by NR method [MVSI]	Index by Machine learning [Exponential GPR method]
19.	86-87	0.0608	0.0608	0.0636	0.0632	0.0858	0.0855	0.0887	0.0859
20.	80-96	0.1301	0.1296	0.0981	0.0815	0.1659	0.1516	0.1686	0.1517
21.	80-99	0.0780	0.0789	0.0703	0.079	0.1871	0.1546	0.1984	0.1926
22.	94-100	0.0593	0.0593	0.07	0.0709	0.0732	0.0677	0.0773	0.0771
23.	80-97	0.0796	0.0796	0.0583	0.0641	0.1051	0.1037	0.1075	0.1067
24.	80-98	0.0534	0.0511	0.0343	0.0194	0.1225	0.0977	0.1282	0.1189
25.	105-106	0.0124	0.0126	0.0117	0.0144	0.0131	0.0148	0.0134	0.0137
26.	32-114	0.0136	0.0134	0.0084	0.0089	0.0177	0.0179	0.0181	0.0196

References

[1] Ohno, T., and Imai, S. 2006. The 1987 Tokyo blackout. *IEEE PES Power systems Conference and Exposition*, Atlanta, GA, USA, 314-318.

[2] Final report on the august 14, 2003, *Blackout in the United States and Canada, Causes and recommendations*, Available, online : https://www.osti.gov/etdeweb/biblio/20461178(2019).

[3] Ajjarapu, V., and Christy, C. 1992. The continuous power flow: A Tool for steady state voltage stability analysis. *IEEE Transaction on Power Systems*, 416-423.

[4] Chiang, H. D., and Jean-Jumeau, R. 1995. Toward a Practical Performance Index for Predicting Voltage Collapse in Electric Power Systems. *IEEE Trans. Power Syst*. 584–592.

[5] Moghavemi, M., and Omar, F. M. 1998. Technique for contingency monitoring and voltage collapse prediction. *IEEE proceedings of Generation Transmission Distribution*, 634-40.

[6] Musirin, I., and Rahman, T. K. A. 2002. On line voltage stability based contingency ranking using Fast voltage stability index (FVSI). *IEEE/PES Transmission Distribution and exhibition conference*, 1118-23.

[7] Gu, W., and Wan, Q. 2010. Linearized voltage stability index for wide area voltage monitoring and control. *International Journal of Electrical Power and Energy Systems*, 333-336.

[8] Devaraj, D. and Roselyn, J. P. 2011. Online voltage stability assessment using radial basis function network model with reduced input features. *International Journal of Electrical Power and Energy systems*, 1550-5.

[9] Rajive Tiwari, K. R., and Nicci Vikas Gupta. 2012. Line collapse proximity index for prediction of voltage collapse in power systems. *International Journal of Electrical Power and Energy Systems*, 05-111.

[10] Kanimozhi, R., and Selvi, K. 2013. A novel Stability index for voltage stability analysis and contingency ranking in power systems using Fuzzy based load flow, *Journal of Electrical Engineering and Technology*, 694-703.

[11] Perez-londono, S., Rodriguez, L. F., and Olivar, G. A. 2014. Simplified voltage stability index (SVSI). *Electr. Power Energy Syst*, 806–813.

[12] Mishra, A., Gundavarapu, V. N., Bathina, V. R., and Duvvada, D. C. 2016. Real power performance index and line stability index-based management of contingency using firefly algorithm. *IET Generation, Transmission & Distribution*, 2327-35.

[13] Su, H. Y., Chen, Y. C., and Hsu, Y. L. 2016. A synchrophasor based optimal voltage control scheme with successive voltage stability margin improvement. *Appl. Sci.*, 6, 14.

[14] Chen, H., Jiang, T., Yuan, H., Jia, H., Bai, L., and Li, F. 2017. Wide-area measurement-based voltage stability sensitivity and its application in voltage control. *Int. J. Electr. Power Energy Syst*, 87–98.

[15] Saurabh Ratra., Rajive Tiwari., and Niazi, K. R. 2018. Voltage stability assessment in power systems using line voltage stability index, *International journal of Computers and Electrical Engineering*, 1–13.

[16] Li, H., Zhou, L., Mao, M., and Zhang, Q. 2019. Three-layer voltage/Var control strategy for PV clusters considering steady-state voltage stability. *J. Clean. Prod*, 56–58.

[17] Danish, M. S. S., Senjyu, T., Sabory, N. R. K. and Mandal, P. 2019. A Recap of Voltage Stability Indices in the Past Three Decades. *Energies*, 1544-1553.

[18] Nian Zhang., Jiang Xiong., Jing Zhong., and Keenan Leatham. 2018. Gaussian Process Regression method for classification for High-dimensional data with limited samples, *Eighth International Conference on Information Science and Technology (ICIST)*, 2573-3311.

[19] Wehenkel, L. 1995. *Machine Learning Approaches to Power System Security Assessment*. Dissertation, University of Liege.

[20] Wehenkel, L., Van Cutsem, T., Pavella, M., Jacquemart, Y., Heilbronn, B., and Pruvot, P. 1994. Machine learning, neural networks and statistical pattern recognition for voltage security: a comparative study, *Engineering Intelligent Systems for Electrical Engineering and Communications*, 233-245.

[21] Chiang, H. D., Flueck, A. J., Shah, K. S., and Balu, N. 1995. CPFLOW: A practical tool for tracing power system steady state stationary behaviour due to load and generation variations. *IEEE Transaction on power systems*, 623-634.

[22] Katsanevakis, M., Stewart, R. A., and Junwei, L. 2019. A novel voltage stability and quality index demonstrated on a low voltage distribution network with multifunctional energy storage systems. *Electr. Power Syst. Res*, 264–282.

[23] Villa-Acevedo, W. M., Lopez-Lezama, J. M., and Colome, D. G. 2020. Voltage Stability Margin Index Estimation Using a Hybrid Kernel Extreme Learning Machine Approach. *Energies*, 857.

[24] Rajalakshmi, P., and Dr. Rathinakumar, M. 2016. Voltage stability assessment of power system by comparing line stability indices, *International Journal of Advanced Research Trends in Engineering and Technology (IJARTET)*, Special Issue 19.

[25] Goha, H. H., Chuaa, Q. S., Leea, S. W., Koka, B. C., Goha, K. C., and Teob, K. T. K. 2015. Evaluation for Voltage Stability Indices in Power System Using Artificial Neural Network, *International Conference on Sustainable Design, Engineering and Construction, Science Direct, Procedia Engineering*, 1127 – 1136.

[26] Ismail, N. A. M., Zin, A. A. M., Khairuddin, A., and Khokhar, S. 2014. A comparison of voltage stability indices, *IEEE 8th International Power Engineering and Optimization Conference*, 2422-6.

[27] Lim, Z. J., Mustafa, M. W., and Muda, Z. 2012. Evaluation of effectiveness of voltage stability indices on different loadings, *IEEE International Power Engineering and Optimization conference*. DOI: 10.1109/PEOCO.2012.6230925.

[28] Ponnada, A., and Rammohan, N. 2012. A study on voltage collapse phenomenon by using voltage collapse proximity indicators, *International journal of Engineering, Research and Development*, 31-35.

[29] Mohamed, A., Jasmon G. B., and Yusoff, S. 1998. A static voltage collapse indicator using line stability factor, *Journal of Industrial Technology*, 73-85.

[30] Moghavemi, M., and Faruque, O. 1998. Real time contingency evaluation and ranking technique, *IEEE proceedings, Generation, Transmission, Distribution*. 145 (5):517 - 524

[31] Zhao, Y., Chen, J., and Poor, H. V. 2020. A Learning-to-Infer Method for Real-Time Power Grid Multi-Line Outage Identification". *IEEE Trans. Smart Grid*, 555–564.

[32] Farsi, B., Amayri, M., Bouguila, N., and Eicker, U. 2021. On Short-Term Load Forecasting Using Machine Learning Techniques and a Novel Parallel Deep LSTM-CNN Approach". *IEEE Access*, 31191–31212.

[33] Dhibi, K., Fezai, R., Mansouri, M., Trabelsi, M., Kouadri, A., Bouzara, K., Nounou, H., and Nounou, M. 2020. Reduced Kernel Random Forest Technique for Fault Detection and Classification in Grid-Tied PV Systems. *IEEE J. Photovolt.*, 1864–1871.

[34] Wen, S., Zhang, C., Lan, H., Xu, Y., Tang, Y., and Huang, Y. 2021. A hybrid ensemble model for interval prediction of solar power output in ship onboard power systems,. *IEEE Trans. Sustain. Energy*, 14–24.

[35] Lei, X., Yang, Z., Yu, J., Zhao, J., Gao, Q., and Yu, H. 2021. Data-Driven Optimal Power Flow: A Physics-Informed Machine Learning Approach. *IEEE Trans. Power Syst*, 346–354.

[36] Zhang, Y., Wang, X., Wang, J., Zhang, Y., and Deep 2021. Reinforcement Learning Based Volt-VAR Optimization in Smart Distribution Systems. *IEEE Trans. Smart Grid*, 12.

[37] Li, N., Li, B., and Gao, L. 2020. Transient Stability Assessment of Power System Based on XGBoost and Factorization Machine. *IEEE Access*, 28403–28414.

[38] Hong, G., and Kim, Y. S. 2020. Supervised Learning Approach for State Estimation of Unmeasured Points of Distribution Network. *IEEE Access*, 113918–113931.

Biographical Sketches

G. Sandhya Rani working as Assistant Professor in Department of Electrical and Electronics Engineering, Vasavi College of Engineering, Hyderabad, India. She received her BTech in Electrical and Electronics engineering from Achrarya Nagarguna University, India, and MTech with a specialisation in Electrical Power Engineering from Jawaharlal Nehru Technological

University Hyderabad, India. She is now pursuing her PhD in Osmania University, Hyderabad. She has 20 years of teaching experience in various institutes. Her research interest includes Neural Networks and its application to Power systems, Machine learning and IOT, Microprocessors and Microcontrollers.

Muktevi Chakravarthy is working as a Professor and Head of the Department of EEE, Vasavi College of Engineering, Hyderabad, India. He obtained his PhD degree from Jawaharlal Nehru Technological University Hyderabad, India in 2013. Obtained his MTech degree in power systems from Jawaharlal Nehru Technological University Kakinada, India in 2005. Obtained his BTech degree from Nagarjuna University Guntur, India in 1999. He has 18 years of experience in Teaching & Research and He provided consultancy to M/s NR Bearings Pvt. Ltd. on Automation of Cage Brightening Station. His areas of research include power system monitoring and protection, smart grids, hybrid vehicles, solar power MPPT and development of hardware and software for microprocessor /microcontroller applications.

Dr. B. Mangu Working as Professor in Electrical Engineering, University College of Engineering Osmania University, Hyderabad, India. Dr Mangu obtained his BE degree in Electrical Engineering from Osmania University in 2000 and ME degree with a specialisation of Industrial drives and control from Osmania University, Hyderabad, in 2002 PhD from IIT, Bombay in 2016. His areas of research are non-conventional energy (solar PV, wind), Power conditioning, maximum power point tracking, stand-alone and grid connected systems, design of converters for renewable sources integrations, intelligent control of power electronic systems: DSP based control. He is Member of IEEE Power Electronics Society, IEEE Power & Energy Society, IEEE Industrial Electronics Society and IEEE Industry Applications Society. Member of Engineering and Scientific Research (ESR) Groups. He published 15 papers in reputed journals and 28 papers in various National and International conferences. Presently Dr. Mangu is guiding five students for their doctoral degree. He guided more than 25 students for ME dissertation work and more than 20 BE Project works. He served Osmania University at various administrative levels like head of the department, chairperson board of studies, Hostel warden and presently he is serving as Director SC-ST Cell, Osmania University. He delivered more than 20 guest lectures and acted as session chair for international conferences

Editors' Contact Information

Dr. Yogesh Kumar Chauhan
Associate Professor
Department of Electrical Engineering,
Kamla Nehru Institute of Technology
Sultanpur, Uttar Pradesh 228118, India
Email: yogeshchauhan@knit.ac.in

Dr. Ranjan Kumar Behera
Department of Electrical Engineering,
Indian Institute of Technology
Patna Bihta, Patna 801106, Bihar, India
Email: rkb@iitp.ac.in

Dr. Asheesh K. Singh
Professor
M. N. National Institute of Technology Allahabad
Prayagraj 211004, India
Email: asheesh@mnnit.ac.in

Index

F

G

H

I

K

L

M

N

O

P

R

S

T

V

W